# METHODS IN MOLECULAR BIOLOGY



For further volumes:
http://www.springer.com/series/7651

For over 35 years, biological scientists have come to rely on the research protocols and methodologies in the critically acclaimed *Methods in Molecular Biology* series. The series was the first to introduce the step-by-step protocols approach that has become the standard in all biomedical protocol publishing. Each protocol is provided in readily-reproducible step-by-step fashion, opening with an introductory overview, a list of the materials and reagents needed to complete the experiment, and followed by a detailed procedure that is supported with a helpful notes section offering tips and tricks of the trade as well as troubleshooting advice. These hallmark features were introduced by series editor Dr. John Walker and constitute the key ingredient in each and every volume of the *Methods in Molecular Biology* series. Tested and trusted, comprehensive and reliable, all protocols from the series are indexed in PubMed.

# Kupffer Cells

## Methods and Protocols

Edited by

**Myriam Aouadi and Valerio Azzimato**

*Department of Medicine, Integrated Cardio Metabolic Center, Karolinska Institutet, Huddinge, Sweden*

Humana Press

*Editors*
Myriam Aouadi
Department of Medicine
Integrated Cardio Metabolic Center
Karolinska Institutet
Huddinge, Sweden

Valerio Azzimato
Department of Medicine
Integrated Cardio Metabolic Center
Karolinska Institutet
Huddinge, Sweden

ISSN 1064-3745 ISSN 1940-6029 (electronic)
Methods in Molecular Biology
ISBN 978-1-0716-0703-9 ISBN 978-1-0716-0704-6 (eBook)
https://doi.org/10.1007/978-1-0716-0704-6



This Humana imprint is published by the registered company Springer Science+Business Media, LLC part of Springer Nature.
The registered company address is: 1 New York Plaza, New York, NY 10004, U.S.A.

# Preface

This book provides readers with a complete overview of kupffer cell biology, from isolation to phenotyping. Isolation can be performed using different methods and requires different protocols when working with mouse and human livers or liver xenograft, hence the different chapters listed. Most of these chapters refer to mouse kupffer cell techniques. Chapters regarding phenotyping include RNA and DNA sequencing as well as lipid profiling, western blot techniques, flow cytometry, and microscopy. We have also included techniques to manipulate gene expression in vitro and in vivo. Finally, we have added chapters regarding disease models from fatty liver disease to cancer.

*Huddinge, Sweden*

*Myriam Aouadi*
*Valerio Azzimato*

# Contents

# Contributors

OSMAN AHMED • *Division of Clinical Chemistry, Department of Laboratory Medicine, Karolinska Institutet, Stockholm, Sweden*
MYRIAM AOUADI • *Department of Medicine, Integrated Cardio Metabolic Center, Karolinska Institutet, Huddinge, Sweden*
VALERIO AZZIMATO • *Department of Medicine, Integrated Cardio Metabolic Center, Karolinska Institutet, Huddinge, Sweden*
EMELIE BARREBY • *Department of Medicine, Integrated Cardio Metabolic Center, Karolinska Institutet, Huddinge, Sweden*
ADELINE BERTOLA • *Department of Medicine, University of Massachusetts Medical School, Worcester, MA, USA*
NIKLAS K. BJÖRKSTRÖM • *Department of Medicine Huddinge, Center for Infectious Medicine, ANA Futura, Karolinska Institutet, Karolinska University Hospital Huddinge, Stockholm, Sweden*
CAMILLE BLÉRIOT • *Singapore Immunology Network (SIgN), Agency for Science, Technology and Research (A*STAR), Singapore, Singapore*
JEREMIE BOUCHER • *Bioscience Metabolism, Research and Early Development, Cardiovascular, Renal and Metabolic (CVRM), BioPharmaceuticals R&D, AstraZeneca, Gothenburg, Sweden; The Lundberg Laboratory for Diabetes Research, University of Gothenburg, Gothenburg, Sweden; Wallenberg Centre for Molecular and Translational Medicine, University of Gothenburg, Gothenburg, Sweden*
CHLOÉ BUCH • *Team Pathophysiology of Dyslipidemia, INSERM UMR1231 "Lipids, Nutrition, Cancer" and Université de Bourgogne Franche-Comté, Dijon, France*
YANGFUN CHENG • *Gene Therapy Center, University of Massachusetts Medical School, Worcester, MA, USA; West China School of Medicine, Sichuan University, Chengdu, China*
TERENCE FLOTTE • *Microbiology and Physiologic Systems, Gene Therapy Center, University of Massachusetts Medical School, Worcester, MA, USA*
GWLADYS GERNOUX • *Department of Pediatrics, Gene Therapy Center, University of Massachusetts Medical School, Worcester, MA, USA*
FLORENT GINHOUX • *Singapore Immunology Network (SIgN), Agency for Science, Technology and Research (A*STAR), Singapore, Singapore*
ADRIEN GUILLOT • *Department of Hepatology/Gastroenterology, Charité University Medical Center, Berlin, Germany*
TONY JOURDAN • *Team Pathophysiology of Dyslipidemia, INSERM UMR1231 "Lipids, Nutrition, Cancer" and Université de Bourgogne Franche-Comté, Dijon, France*
MUHAMMAD FARIS BIN MOHD KAIRI • *Singapore Immunology Network (SIgN), Agency for Science, Technology and Research (A*STAR), Singapore, Singapore*
MARION KORACH-ANDRÉ • *Department of Medicine, Metabolism Unit and Integrated Cardio Metabolic Center (ICMC), Karolinska Institute at Karolinska University Hospital Huddinge, Stockholm, Sweden; Clinical Department of Endocrinology, Metabolism and Diabetes, Karolinska University Hospital Huddinge, Stockholm, Sweden*
SHAMIN LI • *Vaccine and Infectious Disease Division, Fred Hutchinson Cancer Research Center, Seattle, WA, USA*

MIRKO E. MINNITI • *Division of Clinical Chemistry, Department of Laboratory Medicine, Karolinska Institutet, Stockholm, Sweden*
CECILIA MORGANTINI • *Department of Medicine, Integrated Cardio Metabolic Center, Karolinska Institutet, Huddinge, Sweden*
CHRISTIAN MUELLER • *Department of Pediatrics, Gene Therapy Center, University of Massachusetts Medical School, Worcester, MA, USA*
EVAN NEWELL • *Singapore Immunology Network (SIgN), Agency for Science, Technology and Research (A*STAR), Singapore, Singapore; Vaccine and Infectious Disease Division, Fred Hutchinson Cancer Research Center, Seattle, WA, USA*
CAROLINA OSES • *Department of Medicine, Integrated Cardio Metabolic Center, Karolinska Institutet, Huddinge, Sweden*
MATTEO PEDRELLI • *Division of Clinical Chemistry, Department of Laboratory Medicine, Karolinska Institutet, Stockholm, Sweden*
OTTO STRAUSS • *Department of Medicine Huddinge, Center for Infectious Medicine, ANA Futura, Karolinska Institutet, Karolinska University Hospital Huddinge, Stockholm, Sweden*
ANDRÉ SULEN • *Department of Medicine, Integrated Cardio Metabolic Center, Karolinska Institutet, Huddinge, Sweden*
QIUSHI TANG • *Gene Therapy Center, University of Massachusetts Medical School, Worcester, MA, USA*
MICHAELA TENCEROVA • *Department of Molecular Physiology of Bone, Institute of Physiology Czech Academy of Sciences, Prague, Czech Republic*
SANTIAGO VERNIA • *MRC London Institute of Medical Sciences, Hammersmith Hospital Campus, London, UK; Institute of Clinical Sciences, Faculty of Medicine, Imperial College London, London, UK*
CONNIE XU • *Department of Medicine, Integrated Cardio Metabolic Center, Karolinska Institutet, Huddinge, Sweden*
SIJIA YU • *MRC London Institute of Medical Sciences, Hammersmith Hospital Campus, London, UK; Institute of Clinical Sciences, Faculty of Medicine, Imperial College London, London, UK*
BADER ZARROUKI • *Bioscience Metabolism, Research and Early Development, Cardiovascular, Renal and Metabolic (CVRM), BioPharmaceuticals R&D, AstraZeneca, Gothenburg, Sweden*

# Chapter 1

# Kupffer Cell and Hepatocyte Isolation from a Single Mouse Liver by Gradient Centrifugation

**Carolina Oses and Myriam Aouadi**

## Abstract

Kupffer cells and hepatocytes maintain liver homeostasis. These cells could be separated based on their size and weight, by centrifugation using a density gradient after a liver perfusion. Here, we describe a methodology to isolate both Kupffer cells and hepatocytes from a single mouse, which provides the unique advantage of studying these two cell types from the same liver.

**Key words** Liver, Kupffer cells, Hepatocytes, Cell isolation, Perfusion, Gradient centrifugation

## 1 Introduction

The liver plays a role in metabolism and immune system function. This organ can be classified into two major populations, parenchymal (hepatocytes) and nonparenchymal cells (endothelial cells, hepatic stellate cells and immune cells) [1].

Hepatocytes represent 60–80% of total liver cells, and their function is more related with metabolism [1]. Hepatocytes have different functions as oxidative metabolism, glycolysis and gluconeogenesis, ureagenesis, β-oxidation of fatty acids, lipogenesis, cholesterol biosynthesis, amino acid breakdown, protein secretion, and bile acid production [2, 3] .

Kupffer cells (KCs) are 20% of the liver nonparenchymal cells, and they represent between 80 and 90% of the total number of macrophages in the body [4]. KCs, the resident macrophages in the liver are located in the sinusoids, supporting hepatocyte function [2]. KCs are in contact with nutrients, microorganisms, and toxic agents from the portal vein through the hepatic system [5]. KCs are essential in maitaining liver homeostasis and function. KCs are involved in iron homeostasis, removing red blood cells–derived vesicles from circulation [6, 7]. KCs are responsible for clearance of foreign materials such bacteria or viruses [8–10] . KCs sense

Myriam Aouadi and Valerio Azzimato (eds.), *Kupffer Cells: Methods and Protocols*, Methods in Molecular Biology, vol. 2164, https://doi.org/10.1007/978-1-0716-0704-6_1, 

exogenous and endogenous molecular signals, as lipopolysaccharide (LPS) and complement C3a and C5a. Some of these components activate KCs, thus playing a role in inflammatory responses [11].

Liver homeostasis is maintained by hepatocytes and KCs, principally, and it is imperative develop techniques that can be used to isolate and study these cells. Multiple protocols has been developed to isolate hepatocytes and KCs, using liver perfusion, enzymatic digestion, and gradient centrifugation [12, 13]. Several limitations have been identified: expensive equipment, low reproducibility, and isolation of only one hepatic cellular fraction.

Here we describe a low-cost and reproducible protocol to isolate hepatocyte and KCs from a single mouse liver. This protocol is based on hepatic perfusion with collagenase, followed by density gradient centrifugation to separate hepatic cells according to their density using low speed.

## 2 Materials

### 2.1 Liver Perfusion and Cell Isolation

#### 2.1.1 Mice Liver Perfusion Reagents

1. 1× HBSS buffer: Freshly prepare 1× HBSS using 10× HBSS and ultrapure water, filter through a 0.22 μm pore size filter.
2. 0.5 mM HBSS–EGTA buffer: 500 μL 0.5 M EGTA and 500 mL 1× HBSS.
3. 1 mM HBSS–$CaCl_2$ buffer: 500 μL of 1 M $CaCl_2$ and 500 mL 1× HBSS (*see* **Note 1**).
4. 1 mM concentrated collagenase: 1.6 g of collagenase from *Clostridium histolyticum*, 50 mL HBSS–$CaCl_2$ buffer. Filter the solution through a 0.45 μm pore size filter and prepare aliquots of 1 mL that are stored at −20 °C. Dilute 1 mL of concentrated collagenase in 49 mL of HBSS–$CaCl_2$ buffer and heat to 37 °C in water bath before use.
5. Mouse anesthesia: 200 μL per mouse of 80 mg/kg of ketamine, 1 mg/kg medetomidine, phosphate buffer solution (PBS).

#### 2.1.2 Density Gradient for KC Isolation Using Percoll

1. 90% Percoll: For a final volume of 50 mL, add 5 mL of 10× PBS to 45 mL of Percoll.
2. 25% Percoll: For a final volume of 40 mL, add 30 mL of 1× PBS to 10 mL of 90% Percoll.
3. 50% Percoll: For a final volume 30 mL, add 15 mL of 1× PBS to 15 mL of 90% Percoll (*see* **Note 2**).
4. Percoll gradient: In a 50 mL Falcon tube, add 20 mL of 25% Percoll and underlay slowly 14.5 mL of 50% Percoll, at the bottom of the tube using a 10 mL plastic pipette (*see* **Note 3**).

#### 2.1.3 Media

1. Adherence medium M199 for hepatocytes: 480 mL of M199 + Earle's Salts, 5 mL of Penicillin/Streptomycin (P/S), 5 mL of BSA (10% final concentration), 10 mL of Ultroser™ G serum substitute (Pall) or fetal bovine serum, 5 μL of 10 mM dexamethasone, and 72.5 μL of 690 μM Insulin.
2. Basal (maintenance) medium for hepatocytes: 495 mL of M199 + Earle's Salts, 5 mL of Penicillin/Streptomycin (P/S), 5 mL of BSA (10% final concentration), 5 μL of 10 mM dexamethasone and 72.5 μL 690 μM of Insulin.
3. RPMI medium for KC cells: 445 mL of RPMI, 50 mL of 10% fetal bovine serum, 5 mL of 1% Penicillin/Streptomycin (P/S).

#### 2.1.4 Immunofluorescence

1. Hepatocyte immunofluorescence: Albumin antibody (R&D Systems) to be used at a concentration of 25 μg/mL.
2. KC immunofluorescence: F4/80 antibody (AbD Serotec) to be used at a concentration of 1:50.
3. Collagen-coated coverslips for hepatocytes: Coverslip, Rat Tail Collagen Type 1 (*see* **Note 4**).

#### 2.1.5 Real-Time PCR Primers

1. Albumin: Forward, TGCTTTTTCCAGGGGTGTGTT.
   Reverse, TTACTTCCTGCACTAATTTGGCA.
2. Clec4f: Forward, GAGGCCGAGCTGAACAGAG.
   Reverse, TGTGAAGCCACCACAAAAAGAG.
3. 36B4: Forward, TCCAGGCTTTGGGCATCA.
   Reverse, CTTT ATCAGCTGCACATCACTCAGA.

## 3 Methods

These procedures should be carried out at room temperature, unless specified otherwise.

### 3.1 Immunofluorescence Antibodies

1. Prepare the albumin antibody (R&D Systems) for the hepatocytes at a concentration of 25 μg/mL.
2. Prepare the F4/80 antibody (AbD Serotec) for the KCs at a concentration of 1:50.
3. Coat the coverslips using Rat Tail Collagen Type 1 (Sigma) for 6 h at room temperature or overnight at 4 °C. Aspirate the Collagen solution and then dry the coverslips under UV light for 3 h (*see* **Note 4**).

### 3.2 Mouse Perfusion

1. Warm the HBSS–EGTA buffer, HBSS–$CaCl_2$ buffer, hepatocyte adherence media M199, and RPMI media at 37 °C in a water bath.

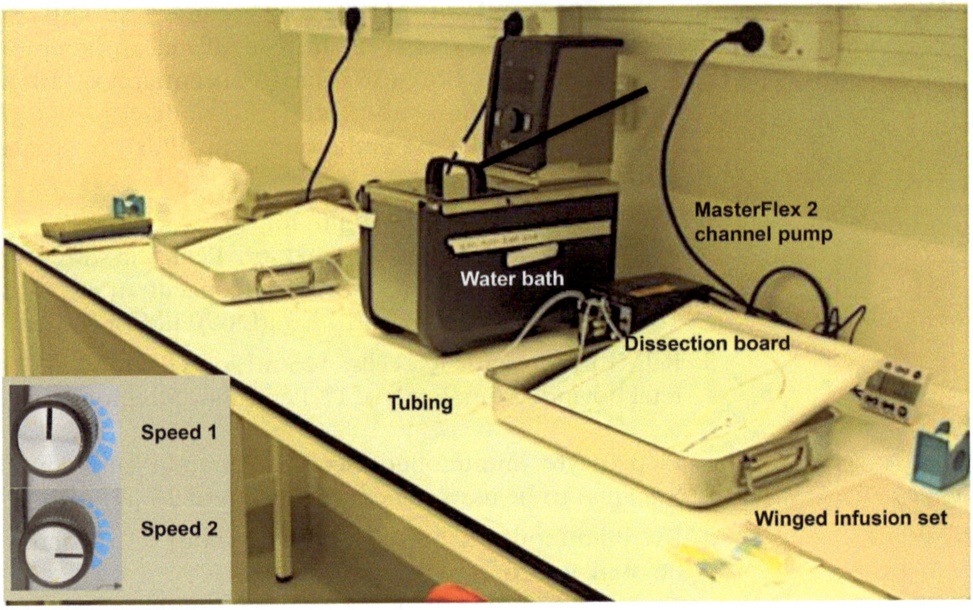


**Fig. 1** Setup for mouse liver perfusion

2. Prepare all instruments including a dissection board, bulldog clamp (World Precision Instruments), and winged infusion set—23 G × 0.75 in. needle × 12 in. tubing (BD Safety-Lok™) and 27 G × 1/2 in. (Terumo Corporation). Connect the Tygon® silicone tubing 1/31″ ID 1/8″ OD (Cole-Parmer) to the winged infusion set and pass it through the MasterFlex® 2 channel Pump (ColeParmer) (Fig. 1).
3. Rinse the tubing with PBS, then with HBSS–EGTA for 2 min.
4. Anesthetize each mouse with ketamine/medetomidine and wait until it no longer demonstrates a withdrawal reflex (*see* **Note 5**).
5. Place one mouse on its back on a dissection board, wipe the abdomen with ethanol, spread the limbs, and secure them to the dissection board.
6. Make an incision through the skin on the ventral midline and dissect the skin away.
7. Turn on the pump and set it to speed 1 (0.38 mL/min) (Fig. 1, **insert**). One end of the tubing should be in the warm HBSS–EGTA and the other should be attached to the winged needle.
8. Open the peritoneal cavity and push the intestines to the right side using a cotton tip; identify both the inferior vena cava and hepatic portal vein (Fig. 2a).

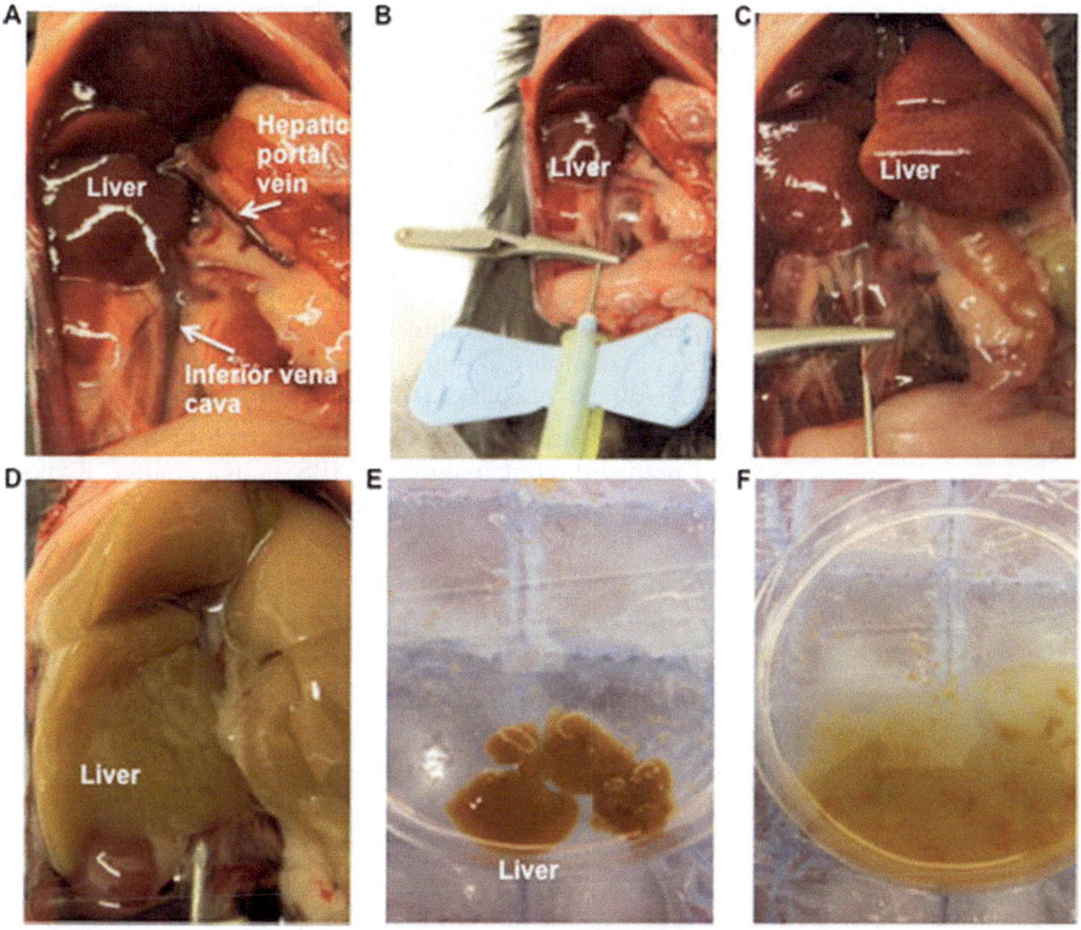


**Fig. 2** Procedure for hepatic cell isolation from single mouse liver. (**a**) Inferior vena cava and hepatic portal vein in mouse abdominal cavity. (**b**) Winged needle inserted in the inferior vena cava and secured with a bulldog clamp. (**c**) Initiation of liver discoloration before sectioning the hepatic portal vein. (**d**) Appearance of liver after collagenase digestion. (**e**) Liver dissected after digestion and placed in a petri dish with cold HBSS–$CaCl_2$ buffer. (**f**) Minced liver

9. Insert the winged needle in the inferior vena cava. Clamp the vena cava with a bulldog clamp in order to secure the needle (Fig. 2b). If the perfusion is working, the liver should swell and become discolored (Fig. 2c).
10. Set the pump to speed 2 (1.7 mL/min) (Fig. 1, **insert**).
11. Cut the hepatic portal vein. This procedure can also be reversed by inserting the needle into the hepatic portal vein and cutting the inferior vena cava instead (*see* **Note 6**).
12. The liver should now discolor rapidly to a creamy color. If the discoloration is slow or red patches remain, the perfusion has not worked, and the liver should not be used for the remaining steps.

13. Once the perfusion of the liver has begun, continue to perfuse the liver with approximately15–20 mL of HBSS–EGTA buffer for approximately 10 min at speed 2 (Fig. 1, **insert**).
14. Stop the pump and transfer the tubing to the collagenase buffer.
15. Perfuse the liver until it is sufficiently digested (Fig. 2d). Signs of sufficient digestion are cell release and lobe breakdown when the liver is pressed with a cotton tip. Digestion of the liver in 10–15 mL of collagenase buffer takes approximately 5–10 min at speed 2 for a lean, healthy mouse. The digestion of the liver for an obese mouse at speed 2 takes approximately 15–30 min and 25–30 mL of collagenase buffer (*see* **Note 7**).
16. After the perfusion is complete, collect the liver and transfer it to a petri dish containing cold HBSS–$CaCl_2$ buffer. Remove the capsule of Glisson while working on ice (Fig. 2e).
17. Release the hepatic cells gently by cutting the liver lobes and continue until all of the big clumps are gone (Fig. 2f).
18. Filter the cells through a 100 μm cell strainer using 30 mL of cold HBSS–$CaCl_2$ buffer and keep the cells on ice.
19. Centrifuge for 3 min at a speed of 50 × *g* at 4 °C.
20. Collect the supernatant containing the KCs. The hepatocytes are in the pellet (Fig. 3) (*see* **Note 8**).

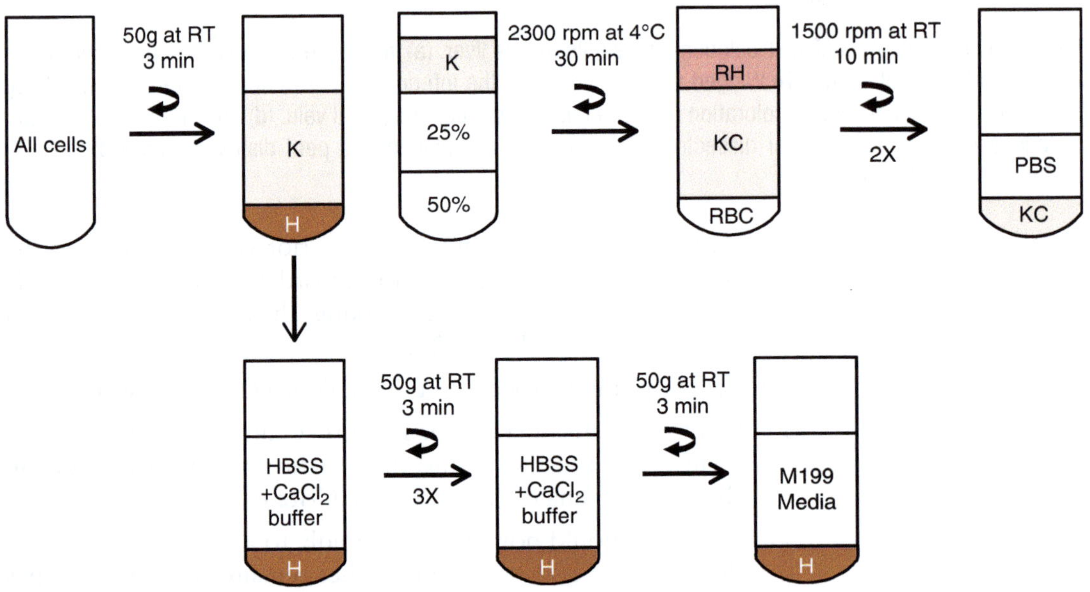


**Fig. 3** Protocol for hepatic cell fraction separation. *K* Kupffer cell (KC)–enriched fraction, *H* hepatocyte-enriched fraction, *RH* rest of hepatocytes, *RBC* red blood cells and cell debris

### 3.3 Hepatocyte Fraction

1. Gently resuspend the hepatocyte pellet in 30 mL of HBSS–$CaCl_2$, by first resuspending the pellet in a smaller volume of 10 mL of HBSS–$CaCl_2$ with a plastic Pasteur pipette, and then gradually adding the remaining 20 mL of the solution. Centrifuge for 3 min at 50 × *g* at 4 °C. Aspirate the rinse buffer supernatant.
2. Repeat this step three times.
3. Resuspend the hepatocytes in 30 mL of adherence media M199. Centrifuge for 3 min at 50 × *g* at 4 °C.
4. Aspirate the supernatant and resuspend the hepatocyte pellet in 10 mL of adherence media M199.
5. Count the cells (*see* **Note 9**). An efficient perfusion should yield approximately $1 \times 10^8$ cells per liver of an adult mouse.
6. Plate the cells for 3–4 h, and then change the adherence media to the maintenance media (*see* **Note 10**).

### 3.4 Kupffer Cell Fraction

1. After the first hepatic cell centrifugation, slowly layer the supernatant on the Percoll gradient with a 10 mL pipette and centrifuge the mixture at 1200 × *g* for 30 min at 4 °C without brakes (*see* **Note 11**).
2. Carefully collect the middle interphase (white cell ring) and transfer it into a new 50 mL tube (*see* Fig. 3b).
3. Wash the cells twice by adding PBS up to a final volume of 20 mL and centrifuging at 1500 rpm for 10 min at 4 °C.
4. Resuspend in KC media (*see* **Note 12**) and plate the cells for 30 min. Wash the cells with PBS and add fresh media to remove the nonadherent cells.
5. The purity of the cell fractions can be observed by microscopy using the immunofluorescent antibodies against the macrophage marker F4/80 and the hepatocyte marker albumin (Fig. 4a, b, *see* **Note 12**). For quantitative analysis, the expression of the KC marker C-Type Lectin Domain Family 4, Member F (Clec4f) [14] and the hepatocyte marker albumin is measured by real-time PCR (Fig. 4c, d).
6. Figure 4 shows that all cells in the KC fraction are stained with F4/80 (Fig. 4a), while all cells in the hepatocyte fraction are positive for albumin (Fig. 4b). Consistent with this observation, Clec4f is highly expressed in the KC fraction (Fig. 4c), while Alb expression is high in hepatocytes (Fig. 4d).

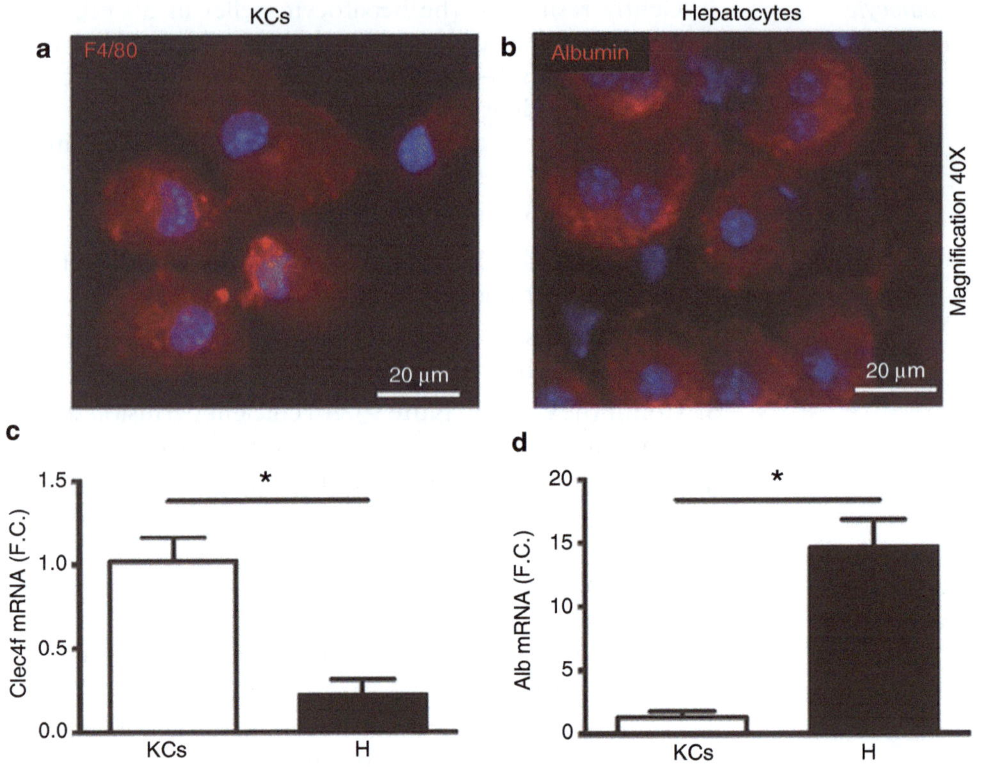


**Fig. 4** Purity of isolated Kupffer cell (KC) and hepatocyte fractions. Cells were isolated from the liver of 14-week-old male C57Bl6/J mice ($n = 3$) on a regular chow diet. (**a**) KCs were stained with an antibody against F4/80 (red) and (**b**) hepatocytes were stained with an antibody against albumin (red). Nuclei were stained with DAPI (blue) (40× magnification). Scale bar, 20 μm. (**c**) Expression of Clec4f and (**d**) albumin (Alb) in KC and hepatocyte (H) fractions; values were normalized using the housekeeping gene 36B4. Results are expressed as the mean fold change (F.C.) value normalized to KC expression ± SEM. Statistical significance was analyzed by Student's t test. $*p < 0.05$

## 4 Notes

1. The 500 mL volume of 1× HBSS–$CaCl_2$ is enough for up to three mice. Half of the volume is heated to 37 °C for the perfusion procedure and the other half is kept on ice for the hepatocyte isolation.
2. When preparing the Percoll solution, calculate the required volume accordingly: 1 tube per 1 lean (WT) mouse; and 2 tubes per obese mouse (>35 g). Calculate 2 tubes per mouse for loading the supernatant suspension from the perfusion procedure.
3. The Percoll gradient needs to be prepared fresh before starting the isolation in order to keep the different layers of the gradient separated.

4. The coverslips can be coated ahead of time, stored at 4 °C, and sealed in Parafilm M®.
5. Lean mice (≈30 g) take about 10–15 min to get anesthetized; for obese mice (>35 g) it takes approximately 30–40 min.
6. During the perfusion, decrease the rate of the pump to speed 1 to see if the winged infusion set is inserted in the correct position in the vein. If the perfusion does not produce a brighter liver, it is recommended to start over with a new mouse. Nevertheless, the perfusion could continue, but the viability and yield of the cells will be lower compared to a liver that was properly perfused.
7. A longer collagenase digestion of the liver will decrease the cell viability.
8. After centrifugation, work with the cell fractions in sterile conditions in a cell culture hood.
9. Good viability of hepatocytes is above 80%.
10. It is recommended to start experiments 3 h after plating.
11. Use caution while loading the Percoll gradient with supernatant. Use a 10 mL Pasteur pipette with the tip touching the wall of Falcon tube.
12. The KC viability is usually between 90 and 100%. The count varies depending on the mice and the quality of perfusion. Approximately 300,000–500,000 cells can be collected from a liver of C57Bl/6J mice on a regular chow diet, while one to two million cells/mouse can be obtained from C57Bl/6J obese mice (>35 g). In addition, flow cytometry analysis shows that 70–80% cells are positive for the F4/80.

## References


1. Gao B, Il Jeong W, Tian Z (2008) Liver: an organ with predominant innate immunity. Hepatology 47(2):729–736
2. Ben-Moshe S, Itzkovitz S (2019) Spatial heterogeneity in the mammalian liver. Nat Rev Gastroenterol Hepatol 16(7):395–410
3. Mason TM (1998) The role of factors that regulate the synthesis and secretion of very-low-density lipoprotein by hepatocytes. Crit Rev Clin Lab Sci 35(6):461–487
4. Duarte N, Coelho IC, Patarrão RS, Almeida JI, Penha-Gonçalves C, Macedo MP (2015) How inflammation impinges on NAFLD: a role for kupffer cells. Biomed Res Int 2015:1–11
5. Ma Y, Yang M, He Z, Wei Q, Li J (2017) The biological function of Kupffer cells in liver disease. In: Biology of myelomonocytic cells, vol 2. InTech, London, p 64
6. Willekens FLA (2005) Liver Kupffer cells rapidly remove red blood cell-derived vesicles from the circulation by scavenger receptors. Blood 105(5):2141–2145
7. Theurl M et al (2008) Kupffer cells modulate iron homeostasis in mice via regulation of hepcidin expression. J Mol Med 86(7):825–835
8. Tu Z, Pierce RH, Kurtis J, Kuroki Y, Crispe IN, Orloff MS (2010) Hepatitis C virus core protein subverts the antiviral activities of human Kupffer cells. Gastroenterology 138 (1):305–314
9. Boltjes A, Movita D, Boonstra A, Woltman AM (2014) The role of Kupffer cells in hepatitis B and hepatitis C virus infections. J Hepatol 61 (3):660–671

10. Surewaard BGJ et al (2016) Identification and treatment of the Staphylococcus aureus reservoir in vivo. J Exp Med 213(7):1141–1151
11. Li P, He K, Li J, Liu Z, Gong J (2017) The role of Kupffer cells in hepatic diseases. Mol Immunol 85:222–229
12. Mito M, Sawa M (1999) Hepatocyte transplantation. Transpl Int 12(6):471–472
13. Ariyoshi K et al (2017) Rapid isolation of murine primary hepatocytes for chromosomal analysis. In Vitro Cell Dev Biol Anim 53 (5):474–478
14. Tencerova M et al (2015) Activated Kupffer cells inhibit insulin sensitivity in obese mice. FASEB J 29(7):2959–2969

# Chapter 2

# Kupffer Cell Isolation from Human Biopsies


## Cecilia Morgantini



## Abstract

Liver macrophages (LMs) are phagocytic cells that play an important role in many liver disorders due to their ability to respond to a variety of stimuli and activating signals.

It is currently debated whether LMs activation from an anti-inflammatory to a proinflammatory phenotype contributes to obesity-induced metabolic diseases. We recently found that LMs can produce noninflammatory factors, such as the protein IGFBP7, able to directly regulate hepatic glucose production and lipid accumulation in the liver. However, while in a mouse model of obesity and insulin resistance LM-*Igfbp7* expression is pathologically increased, in obese insulin-resistant patients LM-*IGFBP7* is edited at RNA level independently of an effect on its expression. This discrepancy between results in animals and humans confirms the importance to perform molecular investigation directly on human's isolated cells. Here, we describe a protocol to isolate liver macrophages from human liver biopsy.



**Key words** Liver biopsy, Humans, Macrophages, Diabetes, Obesity, RNA


## 1 Introduction

Obesity-associated insulin resistance is one of the major risk factors for the development of liver steatosis [1]. Liver steatosis is characterized by the accumulation of lipid in the liver and it is emerging as the most common cause of chronic liver disease worldwide. We recently found that liver macrophages (LMs) can produce noninflammatory factors, such as the protein IGFBP7, able to directly regulate hepatic glucose production and lipid accumulation in the liver through binding to the insulin receptor. In a mouse model of obesity and insulin resistance LM-*Igfbp7* expression is pathologically increased. However, in obese insulin-resistant patients, *IGFBP7* was subjected to RNA editing, a process by which RNA bases can be switched, leading to a new amino acid in the corresponding protein sequence. We found that a particular RNA editing event leading to a new IGFBP7 isoform occurred in the majority of obese insulin-resistant patients. This new isoform had a higher capacity to bind to the insulin receptor and activate AKT compared to wt IGFBP7 in insulin-resistant liver organoids [2].


Myriam Aouadi and Valerio Azzimato (eds.), *Kupffer Cells: Methods and Protocols*, Methods in Molecular Biology, vol. 2164, https://doi.org/10.1007/978-1-0716-0704-6_2, © Springer Science+Business Media, LLC, part of Springer Nature 2020

These results, showing an important indifference in mouse and human, underline the importance to perform molecular analysis on human cells in order to directly relate the finding to human physiology.

Here, we provide a detailed protocol to isolate liver macrophages from a human biopsy.

## 2 Materials

### 2.1 Liver Tissue

A small piece of liver tissue is freshly collected by the surgeon from volunteers undergoing a gastric surgery or obtained by a needle biopsy. In the latter, a small needle is inserted through the abdominal skin into the liver to collect a tissue sample.

### 2.2 Enzymes for Tissue Digestion

1. Collagenase II (sigma C6885): prepare a stock concentration of 50 mg/mL in RPMI media (i.e., dissolve 0.5 g in 50 mL of RPMI media), filter through a 0.22 μm filter and make aliquot of 500 μL. Store aliquots at −20 °C.
2. DNase I (Roche 10,104,159,001): prepare a stock concentration of 10 mg/mL in RPMI media, filter through a 0.22 μm filter, and make aliquots of 500 μL. Store aliquots at −20 °C.

### 2.3 Media

1. Media: Sterile RPMI media (ThermoFisher cat # 21875034).
2. Sterile complete media (RPMI media +10% FCS + 0.5% pen/strep).

### 2.4 Preparation of Percoll Gradient

1. 90% Percoll: For a final volume of 50 mL, add 5 mL of 10× PBS to 45 mL of Percoll (GE Healthcare cat # 15-0891-01).
2. 25% Percoll: For a final volume of 40 mL, add 30 mL of 1× PBS to 10 mL of 90% Percoll.
3. 50% Percoll: For a final volume 30 mL, add 15 mL of 1× PBS to 15 mL of 90% Percoll.
4. Percoll gradient: In a 50 mL Falcon tube, add 20 mL of 25% Percoll and underlay slowly 14.5 mL of 50% Percoll, at the bottom of the tube using a 10 mL plastic pipette.

### 2.5 Other

1. Sterile scalpel.

## 3 Methods

### 3.1 Cell Isolation from a Liver Biopsy

Before staring the procedure, thaw DNase and collagenase on ice, warm up the water bath to 37 °C, and bring media and PBS to room temperature.

1. Transfer the liver tissue into a petri dish.

2. Cover the tissue with 1 mL (*see* **Note 1**) of RPMI media.
3. Cut into smaller pieces (using a scalpel).
4. Transfer the pieces and media into 50 mL tube then add 1 mL of RPMI media (total media is now 2 mL, *see* **Note 1**).
5. Add collagenase: 25 μL/mL *(=50 μL for 2 mL total volume) (stock concentration = 10 mg/mL. Final concentration ≈ 0.25 mg/mL).*
6. Add DNase 20 μL/mL *(=40 μL for 2 mL total volume) Stock concentration = 10 mg/mL. Final concentration ≈ 0.2 mg/mL).*
7. Transfer the tube into water bath ≈37 °C for 30 min with fast shaking.
8. Take out of the tube from the water bath and add 10 mL of complete media (containing FBS) to stop the enzymatic digestion.
9. Filter by passing the cell suspension through the cell strainer (100 or 75 μm). Keep the cell suspension on ice.
10. Centrifuge for 3 min at 50 × *g* at 4 °C.
11. Keep the supernatant that contains liver macrophages that will be used for the Percoll gradient separation.

### 3.2 Density Gradient Isolation of Liver Macrophages Using Percoll

1. Slowly layer the supernatant on the Percoll gradient with a 10 mL pipette and spin down at 1500 × *g* for 30 min without brakes.
2. Carefully collect the middle interphase (white cell ring) and transfer it into a new 50 mL tube.
3. Wash the cells twice by adding PBS up to a final volume of 20 mL and centrifuging at 640 × *g* for 10 min at 4 °C.
4. Resuspend in RPMI complete media and plate the cells for 30 min. Wash the cells with PBS and add fresh media to remove the nonadherent cells.
5. Harvest the cell for further analysis.

## 4 Notes

1. These volumes are calculated based on liver border biopsy (usually taken during abdominal surgery). If the starting material is from a needle biopsy, volumes need to be scaled down.

### References

1. Bellentani S (2017) The epidemiology of non-alcoholic fatty liver disease. Liver Int 37 (Suppl 1):81–84. https://doi.org/10.1111/liv.13299
2. Morgantini C, Jager J, Li X et al (2019) Liver macrophages regulate metabolism through non-inflammatory factors. Nat Metab 1:445–459

# Chapter 3

# Liver Macrophage Isolation by Flow Cytometry Sorting

André Sulen

## Abstract

Liver macrophages are immune cells that under steady state conditions play important roles in maintaining tissue homeostasis. But these cells are also intricately linked to the pathogenesis of obesity-related diseases such as nonalcoholic fatty liver disease (NAFLD), nonalcoholic steatohepatitis (NASH), and type 2 diabetes (T2D). Described herein is a protocol to isolate liver macrophages by FACS for downstream molecular analysis. Such studies of macrophages are important for elucidating their role in the pathogenesis of these metabolic diseases.

**Key words** FACS, Isolation, Insulin sensitivity, Macrophages, Liver, Obesity, Diabetes, Inflammation, Innate immunity

## 1 Introduction

The liver is a heterogeneous tissue composed of parenchymal cells (hepatocytes) and nonparenchymal cells (mostly endothelial cells and immune cells). Liver macrophages are a heterogeneous group of immune cells that in mice have different ontogenies. Kupffer cells have a yolk-sac origin and are seeded during embryogenesis [1], while in the postnatal state, monocyte-derived cells with a hematopoietic origin can obtain Kupffer cell characteristics upon niche availability [2]. Liver macrophages are important for carrying out key liver functions, such as clearing of senescent erythrocytes [3], pathogens [4] and immune complexes [5], as well as reverse cholesterol transport [6]. This function is crucial in maintaining liver and blood homeostasis. However, in obesity, these cells can also contribute to the development and manifestation of diseases such as NAFLD and NASH [7]. Portal macrophage accumulation occurs in early stages of liver steatosis prior to detection of proinflammatory cytokines [8]. Later stages are, however, associated with inflammation and fibrosis; the latter suggested associated

Myriam Aouadi and Valerio Azzimato (eds.), *Kupffer Cells: Methods and Protocols*, Methods in Molecular Biology, vol. 2164, https://doi.org/10.1007/978-1-0716-0704-6_3, 

with alternatively activated macrophages [7]. Here we provide a detailed protocol for the process of sorting liver macrophages by flow cytometry to study the spectrum of roles they play in liver disease.

## 2 Materials

### 2.1 Cells

1. Liver macrophages are sorted from nonparenchymal cells (NPCs). The current protocol is based on NPCs acquired by perfusion of mice livers according to the method by Aparicio-Vergara et al. [9] (*see* **Note 1**).

### 2.2 Antibody Staining of Cells

1. Staining buffer: Add 2 g bovine serum albumin (BSA) and 800 μL 500 mM ethylenediaminetetraacetic acid (EDTA) to 200 mL phosphate buffered saline (PBS) (Final concentration 1% BSA and 2 mM EDTA). Filter the solution with a 0.22 μM vacuum filter into a storage bottle. Adjust pH to 7.2 and store at 4 °C (*see* **Notes 2** and **3**).
2. Sorting buffer: Dilute staining buffer in PBS with 2 mM EDTA to achieve 0.5% BSA. Store at 4 °C.
3. Purified rat anti-mouse CD16/CD32 (Mouse BD Fc Block™) (*see* **Note 4**).
4. Antibodies: CD45 anti-mouse PE-CF594, F4/80 anti-mouse APC (reconstituted in 100 μL PBS upon delivery), CD11b anti-mouse PE-Cy7 (Table 1).
5. Viability stain: SYTOX Blue diluted 1:10 in PBS.
6. FACS tubes.
7. Spectra Mesh/Nylon Filter, mesh opening 200 μm.

### 2.3 FACS

1. The current protocol is based on using a BD FACSAria Fusion with a 100 μM nozzle.

**Table 1**
**Antibodies for liver macrophage identification**

| Antibody | Fluorophore | Dilution | Vendor | Cat. no. |
|---|---|---|---|---|
| CD45 | PE-CF594 | 1:1000 | BD | 562,420 |
| F4/80 | APC | 1:50 | Bio-Rad | MCA497APC |
| CD11b | PE-Cy7 | 1:200 | BD | 552,850 |

## 3 Methods

### *3.1 Cells*

1. Centrifuge NPCs from perfused mice at 1200 rpm (300 rcf) for 7 min at 4 °C. Discard supernatant before dissolving the pellet by repeatedly flicking the tube. Resuspend cells at $1 \times 10^7$ cells/mL in staining buffer and place on ice. All steps hereafter should be carried out on ice.

### *3.2 Antibody Staining of Cells*

1. Transfer isolated NPCs in 500 μL staining buffer to FACS tubes as described in Table 2. Add 1 μL purified rat anti-mouse CD16/CD32 to each tube (1:500 final dilution), incubate on ice for 10 min.
2. Centrifuge at 1200 rpm (300 rcf) for 5 min at 4 °C. Discard supernatant and dissolve pellet in remaining buffer before adjusting to 50 μL with staining buffer.
3. Add 50 μL of antibody cocktail prepared as described in Table 2. Incubate on ice for 20 min protected from light.
4. Add 2 mL of staining buffer to each tube, centrifuge at 1200 rpm (300 rcf) for 5 min at 4 °C, discard supernatant and dissolve pellet. Repeat once.
5. Resuspend cells in 500 or 300 μL sorting buffer for full stain and controls, respectively.

### *3.3 FACS*

1. Start flow cytometer according to routines and quality controls. Use a 100 μM nozzle to ensure good viability of sorted cells.

**Table 2**
**Preparation of antibody cocktails for staining**

| Tube | Cell number | CD45 (1:20) | F4/80 | CD11b (1:4) | Staining buffer | Total |
|---|---|---|---|---|---|---|
| Full stain | $5 \times 10^6$ | 2 | 2 | 2 | 44 | 50 |
| FMO CD45 | $1 \times 10^6$ | 0 | 2 | 2 | 46 | 50 |
| FMO F4/80 | $1 \times 10^6$ | 2 | 0 | 2 | 46 | 50 |
| FMO CD11b | $1 \times 10^6$ | 2 | 2 | 0 | 46 | 50 |
| Single CD45 | $1 \times 10^6$ | 2 | 0 | 0 | 48 | 50 |
| Single F4/80 | $1 \times 10^6$ | 0 | 2 | 0 | 48 | 50 |
| Single CD11b | $1 \times 10^6$ | 0 | 0 | 2 | 48 | 50 |
| Unstained | $1 \times 10^6$ | 0 | 0 | 0 | 50 | 50 |

The table displays volume (μL) used for the different antibodies and the staining buffer. Making diluted stock solutions of CD45 and CD11b facilitates convenient pipetting volumes

2. Add 1 μL SYTOX Blue per single stain and run tubes to adjust PMTs for optimal separation in each channel. Use unstained sample to determine gate for viability stain.
3. Calculate compensation by running single stains when optimal channel settings have been determined.
4. Add 1 μL SYTOX Blue per FMO and run tubes to determine gating for sorting of liver macrophages (for gating strategy *see* Fig. 1a).
5. Prior to sorting the fully stained sample, filter the cells through a mesh filter into a new FACS tube (Spectra Mesh/Nylon Filter, mesh opening 200 μm).
6. Sort cells (purity mask) into a tube with staining buffer under cooling conditions (*see* **Note 5**).
7. Assess purity of sorted cells by running a few events (Fig. 1b).

## 4 Notes

1. The NPC fraction consists of all the cells remaining after removal of hepatocytes by low-speed centrifugation.
2. Fetal bovine serum (FBS) is not used in buffers due to possible contribution of ribonucleic acids (RNA) that may interfere with downstream molecular analysis.
3. BSA can lower the pH of PBS.
4. Blocking Fc-receptors limits unspecific cross-reactivity with antibodies.
5. Cells can be sorted directly into TRIzol if desired.

## Acknowledgments

Valerio Azzimato and Emelie Barreby are acknowledged for providing cells.

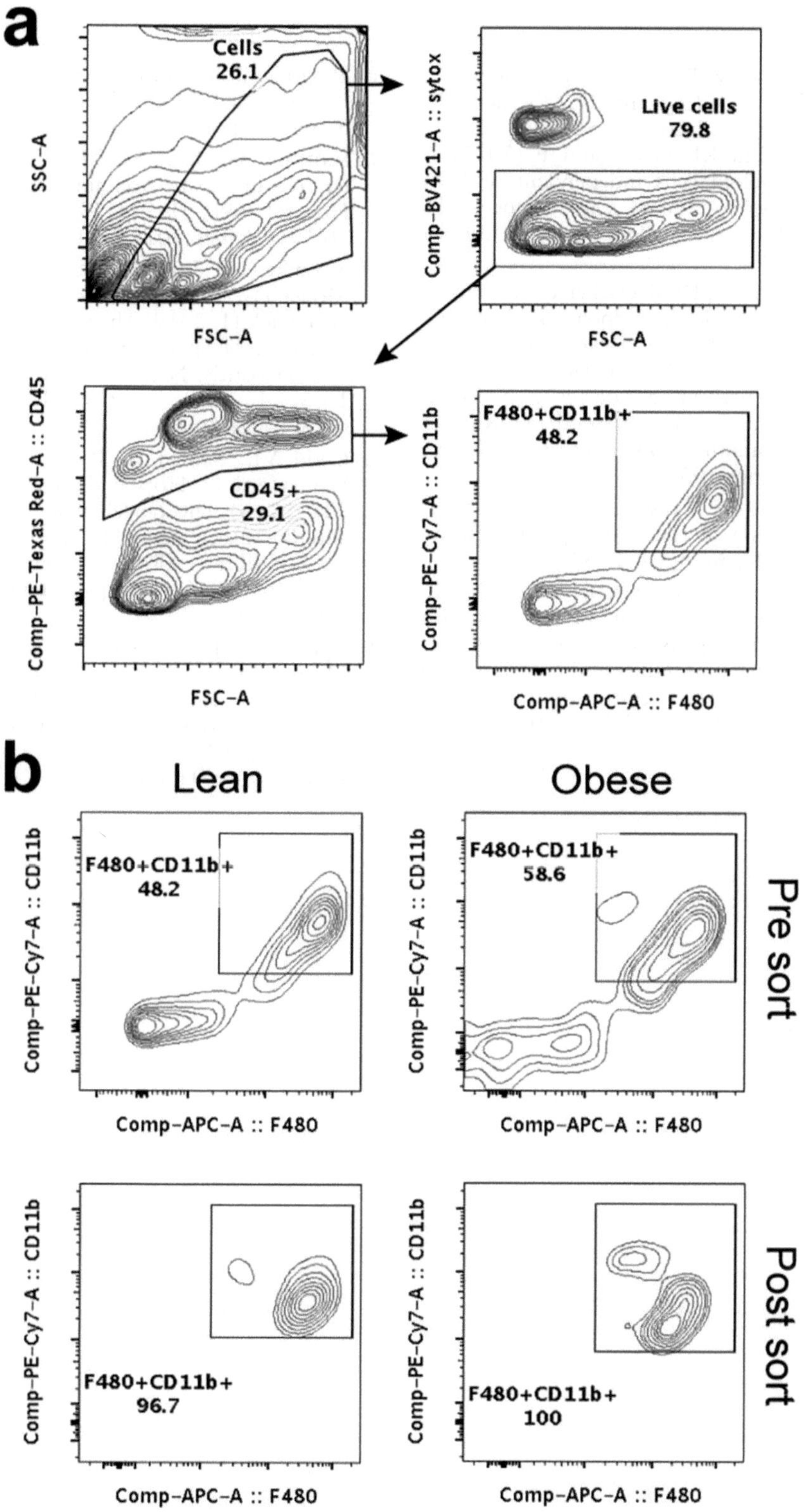


**Fig. 1** Gating strategy and acquired purity when sorting liver macrophages. (**a**) Gating strategy for live CD45$^+$ F4/80$^+$ CD11b$^+$ cells. (**b**) Purity of sorted liver macrophages from lean and obese mice

## References


1. Schulz C, Gomez Perdiguero E, Chorro L, Szabo-Rogers H, Cagnard N, Kierdorf K, Prinz M, Wu B, Jacobsen SE, Pollard JW, Frampton J, Liu KJ, Geissmann F (2012) A lineage of myeloid cells independent of Myb and hematopoietic stem cells. Science 336 (6077):86–90. https://doi.org/10.1126/science.1219179
2. Scott CL, Zheng F, De Baetselier P, Martens L, Saeys Y, De Prijck S, Lippens S, Abels C, Schoonooghe S, Raes G, Devoogdt N, Lambrecht BN, Beschin A, Guilliams M (2016) Bone marrow-derived monocytes give rise to self-renewing and fully differentiated Kupffer cells. Nat Commun 7:10321. https://doi.org/10.1038/ncomms10321
3. Kondo H, Saito K, Grasso JP, Aisen P (1988) Iron metabolism in the erythrophagocytosing Kupffer cell. Hepatology 8(1):32–38
4. Wong CH, Jenne CN, Petri B, Chrobok NL, Kubes P (2013) Nucleation of platelets with blood-borne pathogens on Kupffer cells precedes other innate immunity and contributes to bacterial clearance. Nat Immunol 14 (8):785–792. https://doi.org/10.1038/ni.2631
5. Skogh T, Blomhoff R, Eskild W, Berg T (1985) Hepatic uptake of circulating IgG immune complexes. Immunology 55(4):585–594
6. Scott CL, Guilliams M (2018) The role of Kupffer cells in hepatic iron and lipid metabolism. J Hepatol 69(5):1197–1199. https://doi.org/10.1016/j.jhep.2018.02.013
7. Kazankov K, Jorgensen SMD, Thomsen KL, Moller HJ, Vilstrup H, George J, Schuppan D, Gronbaek H (2019) The role of macrophages in nonalcoholic fatty liver disease and nonalcoholic steatohepatitis. Nat Rev Gastroenterol Hepatol 16(3):145–159. https://doi.org/10.1038/s41575-018-0082-x
8. Gadd VL, Skoien R, Powell EE, Fagan KJ, Winterford C, Horsfall L, Irvine K, Clouston AD (2014) The portal inflammatory infiltrate and ductular reaction in human nonalcoholic fatty liver disease. Hepatology 59 (4):1393–1405. https://doi.org/10.1002/hep.26937
9. Aparicio-Vergara M, Tencerova M, Morgantini C, Barreby E, Aouadi M (2017) Isolation of Kupffer cells and hepatocytes from a single mouse liver. Methods Mol Biol 1639:161–171. https://doi.org/10.1007/978-1-4939-7163-3_16

# Chapter 4

# Kupffer Cell Protein Isolation and Detection by Western Blot

Valerio Azzimato

## Abstract

Kupffer cells are the majority of tissue macrophages present in the body and are actively involved in the regulation of liver homeostasis as well as in the progression of liver diseases. Here we describe a methodology to successfully isolate and detect proteins from Kupffer cells and avoid the contamination of other liver cell populations.

**Key words** Liver, Kupffer cells, Proteins, Protein isolation, Western blot, SDS-PAGE

## 1 Introduction

Kupffer cells (KCs) represent 80–90% of the tissue macrophages present in the body and are increasingly recognized as important modulators which control the liver's response to injury and repair [1]. They play a crucial role in liver homeostasis and are implicated in the initiation, maintenance and outcome of liver inflammation [2, 3]. KCs are localized within the lumen of the liver sinusoids, and they are constantly exposed to several stimuli, such as gut-derived bacteria, microbial debris, and bacterial endotoxins, which eventually could activate macrophages. Upon activation, KCs release various products, including cytokines, nitric oxide, and reactive oxygen species [4, 5]. Therefore, KCs are intimately involved in the liver's response to infection, toxins, ischemia, resection, and other stresses. In addition to that, it has been recently reported that KCs could play a role independent of inflammation [6]. Given the variety of processes in which KCs are involved, it is not surprising that KCs are implicated in the pathogenesis of several liver diseases, ranging from obesity and diabetes, viral hepatitis to steatohepatitis, alcoholic liver disease, intrahepatic cholestasis, activation or rejection of the liver during liver transplantation, and liver fibrosis [7, 8]. For this reason, it has been necessary the development of different techniques which allow the detection of KCs in several conditions. However, despite KCs-related techniques mainly refer to their

Myriam Aouadi and Valerio Azzimato (eds.), *Kupffer Cells: Methods and Protocols*, Methods in Molecular Biology, vol. 2164, https://doi.org/10.1007/978-1-0716-0704-6_4, 

isolation and/or to their transcription regulation [6, 9], a clear protocol for the detection of KCs proteins has never been published.

Here we provide a cost-effective and reproducible protocol for isolation and detection of KCs-specific proteins from a single mouse liver. This method is based on the well-known sodium dodecyl sulfate–polyacrylamide gel electrophoresis (SDS-PAGE) [10].

## 2 Materials

Unless otherwise specified, prepare all solutions using $ddH_2O$ (prepared by purifying deionized water to attain a sensitivity of 18.2 MΩ-cm at 25 °C) and analytical grade reagents, and store all solutions/reagents at 4 °C.

1. KCs previously isolated and plated.
2. 6-well plates.
3. Cell scraper.
4. Phosphate-Buffered Saline (PBS).
5. Lysis Buffer: 2× Laemmli Buffer [11]: 2% (wt/v) SDS, 2% (wt/v) glycerol, Tris–HCl (pH = 7, 65 mM), $H_2O$ (*see* **Note 1**).
6. Protease inhibitor and phosphatase inhibitor cocktails (Roche) (*see* **Note 2**).
7. Reagents or kits for measurement of protein concentration (e.g., bicinchoninic acid (BCA) assay reagents).
8. Spectrophotometer or microplate reader.
9. ThermoBlock.
10. 4–12% SDS-PAGE precast gradient gels (we use NuPage pre-casted gel, Invitrogen).
11. SDS-PAGE running buffer: (20×) NuPAGE™ MES SDS Running Buffer (Invitrogen) (*see* **Note 3**).
12. Protein Loading Buffer (*see* **Note 4**).
13. Protein ladder maker.
14. Protein electrophoresis equipment. We use XCell SureLock® Mini-Cell (Invitrogen).
15. Nitrocellulose membranes.
16. TransBlot® Turbo™ mini nitrocellulose transfer pack.
17. TransBlot® Turbo™ transfer system (Bio-Rad).
18. Ponceau S solution (Roche).
19. TBS containing 0.05% (wt/v) Tween 20 (TBST) (*see* **Note 5**).

20. Blocking solution: 5% (wt/v) skim milk in TBS. Store at 4 °C; (*see* **Note 6**).
21. Primary antibodies. Here we used NRF2 (Abcam) and β-actin (Sigma-Aldrich). Store at −20 °C.
22. Horseradish peroxidase (HRP)-conjugated secondary antibodies specific to the primary antibodies.
23. Chemiluminescence detection reagents (we use Clarity Western ECL Substrate, Bio-Rad).
24. Shaker.
25. Washing trays.
26. Image analysis system: we use ChemiDoc (Bio-Rad).

## 3 Methods

Perform gel electrophoresis and transfer proteins to the membrane according to standard protocols. We employ the semidry transfer method, which is faster than wet transfer and expends less blotting buffer.

### 3.1 Sample Preparation

Prepare protein extract of interest. In this section, we describe our protocol to extract proteins from plated KCs. Carry out all procedures at 4 °C or on ice.

1. Wash cells in the tissue culture dish with cold PBS and gently rock it manually.
2. Discard the PBS. Add ice-cold Laemmli buffer with protease and phosphatase inhibitor cocktails and collect cell lysates with a cell scraper immediately. Pipet the cell lysates into 1.5 mL microcentrifuge tubes.
3. Incubate cell lysates on ice for 30 min. Centrifuge them at 13,000 × *g* for 5 min at 4 °C, and then transfer the supernatants into fresh tubes.
4. Determine the protein concentration of each sample using BCA according to the manufacturer's instructions.
5. Optional: is possible to sonicate cell lysates using to increase the protein yield. This is particularly recommended for heavier proteins or complexes (*see* **Note 7**).
6. Adjust the sample volume to the desired protein concentration to permit loading the same volume for each sample. In our lab, 20–30 μg of proteins are separated for the detection of specific KCs proteins.
7. Add an equal volume of 2× Laemmli sample buffer. Boil the protein samples at 95 °C for 5 min before loading them onto

the gel. After boiling, spin down the samples to bring down the condensation on the inner surface of the tubes.

8. Store at −80 °C before use.

### 3.2 SDS-PAGE

1. Load the samples and protein markers onto the gel (*see* **Note 8**).
2. Run the gel at a constant voltage of 170–180 V for 60 min, or until the dye front runs off the bottom of the gel.

### 3.3 Semidry Electrophoretic Transfer

1. Set up the sandwich for electrotransfer using the transfer kit according to manufacturer's instructions.
2. Insert the membrane gel sandwich into a cassette of the Bio-Rad turbo transfer system.
3. Transfer at 2.5 A (25 V) for 7 min (use Turbo Midi setting).

### 3.4 Probing and Detection

1. Remove the nitrocellulose membrane from the sandwich.
2. Optional: you may want to incubate your membrane in a Ponceau S solution before the blocking step. The Ponceau S stains the proteins and is useful to visualize the quality of the protein transfer and the protein load of each well. Incubate the membranes in a solution of Ponceau S for 5 min under agitation, then rinse briefly several times with water.
3. Incubate membranes for 60 min with blocking buffer at room temperature under gentle agitation.
4. Incubate membrane with primary antibody diluted in blocking buffer overnight at 4 °C under gentle agitation.
5. Proceed to the washing step: add at least 10 mL of TBST 1× for 5 min at room temperature under agitation. Discard the buffer and repeat this step three times.
6. Incubate the membranes in the secondary antibody solution for 1 h at room temperature under gentle agitation. Dilute the appropriate horseradish peroxidase–coupled antibody in the blocking buffer.
7. Wash the membranes with 10 mL of TBST 1× for 5 min at room temperature under agitation. Discard the buffer and repeat this step three times.

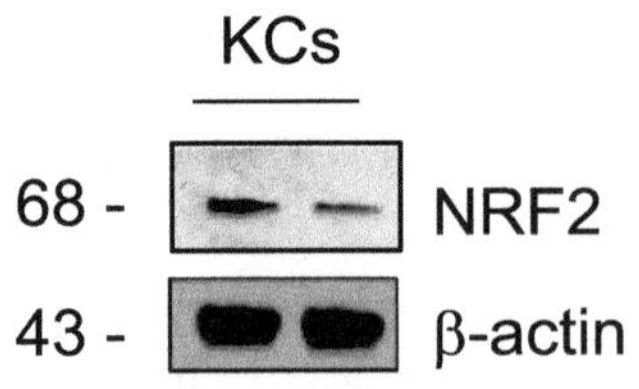


**Fig. 1** Western blot performed on isolated KCs from single mouse liver showing specific signal of NRF2 and β-actin

8. Incubate the membranes in HRP substrate (*see* **Note 9**).
9. Expose the membranes to autoradiography films or to a gel imager. Scan and detect signals (Fig. 1).

## 4 Notes

1. SDS precipitates at 4 °C. Therefore, the lysis buffer needs to be warmed prior to use.
2. To prevent degradation of extracted proteins and to increase the yield of intact proteins, the addition of the proteinase and phosphatase inhibitors to the cell lysis buffer is recommended. We use cOmplete™ EDTA-free Protease Inhibitor Cocktail and PhosSTOP™ Phosphatase Inhibitor Cocktail from Roche.
3. Simple method of preparing 1× running buffer: add 50 mL of (20×) NuPAGE™ MES SDS Running Buffer to a 1 L graduated cylinder and make it to about 950 mL of water.
4. We use Bio-Rad 4× Protein Loading Buffer (Sample buffer). Mix 50 mL $ddH_2O$, 8 mL (pH 6.8) Tris–HCl, 20 mL 10% SDS, 10 mL glycerol, 2.5 mL 2-mercaptoethanol, and ~50 ng of bromophenol blue. Aliquot and store at −20 °C long term and at room temperature after opening.
5. Add 100 mL of 10× TBS to a 1 L graduated cylinder and make it to about 800 mL with water. Add 500 mL of Tween 20 (remember to cut the end of blue tip to aspirate Tween 20 easily).
6. For 150 mL, add 7.5 g nonfat dry milk to 150 ml 1× TBST and mix well.
7. While sonicating remember to wear ear protection. To perform sonication, submerge the sonicator probe into the lysate, and sonicate for 10–15 s at 20–30% of amplitude. Be careful to keep the sonicator probe submerged and avoid the formation of bubbles. Following the 10–15 s sonication, incubate on ice for 30 s before repeating. Repeat this step two or three times.
8. For sample preparation, add appropriate volume of Protein Loading Buffer. We usually load a total volume of 20 μL/sample.
9. We use Bio-Rad Clarity ECL substrate, incubating 1:1 of Solution A and Solution B for 5 min at room temperature in the dark.

## References


1. Duarte N, Coelho IC, Patarrão RS et al (2015) How inflammation impinges on NAFLD: a role for Kupffer cells. Biomed Res Int 2015:984578
2. Kawai T, Akira S (2009) The roles of TLRs, RLRs and NLRs in pathogen recognition. Int Immunol 21:317–337
3. Ramachandran P, Pellicoro A, Vernon MA et al (2012) Differential Ly-6C expression identifies the recruited macrophage phenotype, which orchestrates the regression of murine liver fibrosis. Proc Natl Acad Sci U S A 109: E3186–E3195
4. Tacke F, Zimmermann HW et al (2014) Macrophage heterogeneity in liver injury and fibrosis. J Hepatol 60(5):1090–1096
5. Zimmermann HW, Trautwein C, Tacke F et al (2012) Functional role of monocytes and macrophages for the inflammatory response in acute liver injury. Front Physiol 3:56
6. Morgantini C, Jager J, Li X et al (2019) Liver macrophages regulate metabolism through non-inflammatory factors. Nat Metab 1:445–459
7. Stienstra R, Saudale F, Duval C, Keshtkar S et al (2010) Kupffer cells promote hepatic steatosis via interleukin-1beta-dependent suppression of peroxisome proliferator-activated receptor alpha activity. Hepatology 51 (2):511–522
8. Tencerova M, Aouadi M, Vangala P et al (2015) Activated Kupffer cells inhibit insulin sensitivity in obese mice. FASEB J 29 (7):2959–2969. https://doi.org/10.1096/fj.15-270496
9. Aparicio-Vergara M, Tencerova M, Morgantini C, Barreby E, Aouadi M (2017) Isolation of Kupffer cells and hepatocytes from a single mouse liver. Methods Mol Biol 1639:161–171
10. Borel F, Mueller C (eds) (1639) Alpha-1 antitrypsin deficiency. Methods in molecular biology. Humana Press, New York
11. Laemmli UK (1970) Cleavage of structural proteins during the assembly of the head of bacteriophage T4. Nature 227:680–685

# Chapter 5

# Kupffer Cell mRNA Sequencing

## Emelie Barreby and Connie Xu

## Abstract

The liver is an important organ for the regulation of whole-body metabolism, as well as for immunity. Kupffer cells (KCs) are specialized liver-resident macrophages and the major population of immune cells in the liver. These cells have been shown to play an important role for the regulation of liver homeostasis, and many studies have thus linked these cells to the development of various liver diseases. However, the complexity of macrophage populations and the lack of specific and exclusive markers have so far made it difficult to interpret results from many of these studies. Today, new technologies have emerged including next-generation sequencing allowing for more in depth investigation of multifaceted cell populations such as KCs. Here, we describe a protocol to isolate and prepare cDNA libraries for mRNA sequencing of murine liver macrophages. Using mRNA sequencing to study the gene expression of macrophages in the liver provides a great tool to study the various functions of these cells in the regulation of homeostasis and immunity.

**Key words** Murine Kupffer cells (KCs), RNA extraction, mRNA, mRNA library preparation, Next-generation sequencing, Bioinformatic analysis

## 1 Introduction

The liver has many functions and not only is it crucial for the regulation of whole-body metabolism but is also an immunological organ important for both host defense and immunological tolerance [1, 2]. The majority of cells in the liver are hepatocytes, that make up the liver parenchyma together with cholangiocytes. Aside from these, the liver also contains nonparenchymal cells (NPCs) including liver sinusoidal endothelial cells (LSECs) that line the sinusoidal capillary channels of the liver, and hepatic stellate cells (HSCs) that are located at the space of Disse (the layer between the sinusoids and hepatocytes), together with several immune cell types [3]. Kupffer cells (KCs), specialized liver-resident macrophages, make up the largest population of immune cells in the liver. These cells reside in the liver sinusoids where they serve as important gatekeepers for initiating or suppressing an immune response. The liver is constantly exposed to a large number of antigens,

Myriam Aouadi and Valerio Azzimato (eds.), *Kupffer Cells: Methods and Protocols*, Methods in Molecular Biology, vol. 2164, https://doi.org/10.1007/978-1-0716-0704-6_5, 

microbial products, proteins, and patrolling immune cells coming from the GI tract (through the portal vein) and the hepatic artery [2]. By constantly tasting their environment through phagocytosis, macrophages are equipped to recognize threats to the liver and amplify injury signals, and have therefore been shown to play a central role in the development of multiple of liver diseases [4]. Until recently, the importance of KCs in vivo and their influence on metabolism have mainly been studied by using different transgenic animal models, either by complete knockout models or myeloid-specific knockouts or by depletion of liver myeloid cells by either clodronate liposomes or gadolinium chloride [5]. However, these models are limited in their specificity to KCs and often affect multiple cell types, making it difficult to pinpoint specific functions [6]. Today, the emergence of new technologies including next generation sequencing allows for a deeper understanding of the complex roles of KCs in the regulation of liver homeostasis, without the need for transgenic animal models or depletion studies [7]. Here, we describe a detailed protocol for performing mRNA sequencing of murine liver macrophages. This includes isolation of macrophages from murine livers, extraction of RNA, and preparation of cDNA libraries for mRNA sequencing.

## 2 Materials

### 2.1 Buffers and Solutions for Isolation of Murine Liver Macrophages

Prepare the following buffers and solutions in advance with exception for the Percoll density gradient and collagenase buffer, which need to be freshly prepared on the same day as the liver perfusion. Store buffers, solutions, and media at 4 °C, unless otherwise specified. Solutions are enough for approximately 2 mice.

1. 1× HBSS buffer: Freshly prepare by diluting 10× HBSS 1:10 in ultrapure water.
2. HBSS-EGTA solution (0.5 mM): Add 500 μL of EGTA 0.5 M to 500 mL of 1× HBSS.
3. HBSS-$CaCl_2$ solution (1 mM): Add 500 μL of $CaCl_2$ 1 M to 500 mL of 1× HBSS.
4. Collagenase stock solution (32 mg/mL): Resuspend 1.6 g of collagenase from *Clostridium histolyticum* in 50 mL of HBSS-$CaCl_2$ solution. Filter the solution through a 0.45 pore-sized filter. Prepare 1 mL aliquots and store at −20 °C.
5. Collagenase buffer: Freshly prepare by thawing a 1 mL aliquot of collagenase stock solution and mix with 49 mL of HBSS-$CaCl_2$ solution.
6. Preparation of 25% Percoll solution and 50% Percoll solution for the density gradient isolation: Start by preparing a 90% stock solution by mixing 5 mL of 10× PBS with 45 mL of

Percoll. From the stock solution, take 10 mL and dilute with 30 mL of 1× PBS to make a 25% Percoll solution. To make the 50% Percoll solution, dilute 15 mL of 90% stock solution with 15 mL of 1× PBS.

7. Preparation of Percoll density gradient: Prepare fresh, the same day by adding 20 mL of 25% Percoll into a 50 mL falcon tube. Use a 10 mL serological pipette to add 14.5 mL of 50% Percoll solution at the bottom of the tube by inserting the pipette gently and releasing the 50% Percoll solution slowly, to allow the formation of two separate layers.
8. Media for plating of KCs: Add 50 mL of fetal bovine serum (FBS) and 5 mL of penicillin–streptomycin (P/S) to 500 mL of RPMI.

#### *2.1.1 Liver Perfusion and Cell Isolation*

1. Mouse anesthesia: 200 μL of 100 mg/kg of ketamine and 10 mg/kg xylazine in phosphate buffer solution (PBS) per mouse.
2. Ultrapure 1× PBS.
3. Masterflex 2 channel pump (Cole-Parmer) with tubing.
4. Bulldog clamps.
5. Winged infusion set perfusion needles (23 G).
6. Needles and syringes.
7. 100 μm filters.

#### *2.1.2 Enrichment of Liver Macrophages by Density Gradient*

1. Centrifuge. Make sure that the centrifuge is chilled to 4 °C before use.
2. 6-well culture plates with lid.
3. Ultrapure 1× PBS.
4. Automated cell counter.
5. Cell counting chambers.
6. Trypan blue stain (0.4%).

### 2.2 RNA Extraction of Isolated Liver Macrophages

1. RNaseZap.
2. RNase-free microcentrifuge tubes, 1.5 mL.
3. TRIzol reagent.
4. Chloroform.
5. Isopropanol (propan-2-ol).
6. GlycoBlue coprecipitant.
7. Absolute ethanol.
8. Nuclease-free water.
9. Mini benchtop microcentrifuge.
10. NanoDrop 2000 spectrophotometer.

### 2.3 Library Preparation of Murine Liver Macrophages

1. Single-channel pipettes (20 and 200 μL).
2. 8 or 12-channel multichannel pipette with volume capacity of 200 μL or above.
3. Qubit fluorometer.
4. Qubit High-Sensitivity RNA assay kit.
5. Qubit assay tubes.
6. 2100 Bioanalyzer Instrument.
7. Agilent Bioanalyzer RNA 6000 Nano kit.
8. Agilent Bioanalyzer High-Sensitivity DNA kit.
9. IKA Vortexer with head attachment for Bioanalyzer chip.
10. Illumina TruSeq Stranded mRNA LT Sample Prep Kit (for 48 samples).
11. Illumina TruSeq RNA Single Indexes Set A, 12 Indexes—48 Samples (*see* **Note 1**).
12. Thermal Cycler with 96-Well head attachment.
13. Hard-shell 0.2 mL 96-well PCR plate, semiskirted.
14. Microseal 'B' Adhesive PCR Plate seal.
15. SuperScript II Reverse Transcriptase.
16. Agencourt AMPure XP beads.
17. 80% ethanol (EtOH): Freshly prepare 80% EtOH by diluting absolute ethanol in nuclease-free water.
18. 96-well magnetic stand.
19. Illumina Sequencing Library qPCR Quantification Kit.

### 2.4 Next-Generation Sequencing of Murine Kupffer Cells

1. Illumina sequencing instrument, such as the Illumina HiSeq 3000.

## 3 Methods

### 3.1 Isolation of Murine Liver Macrophages

#### 3.1.1 Liver Perfusion and Cell Isolation

1. Start by warming up the cell culture media and the HBSS-EGTA solution. Prepare the dissection platform by placing a dissection board into a metal basin. Take out the clamps and the instruments and connect the perfusion needles to the tubings of the pump. Rinse the pump with ethanol, PBS, then finally with HBSS-EGTA solution for about 10 min (perfusion rate should be around 5 mL/min without bubbles, make sure that the tubing reaches the bottom of the bottle).
2. Anesthetize mice with ketamine–xylazine mix; wait until animals do not demonstrate a withdrawal reflex.

3. Install 1–2 mice on the dissection board and wipe the fur with alcohol.
4. Open the abdomen, cutting the skin along the midline to the level of the sternum (be careful not to touch the liver), make lateral cuts in the lower abdomen before the midline cut to allow the liver to detach from the abdominal wall more easily.
5. Push the intestines to the side using a cotton tip, identify the interior vena cava and hepatic portal vein.
6. Prime the pump with warm HBSS-EGTA solution, insert the needle into the inferior vena cava (inferior to the liver) keeping the needle flat against the body (see if the liver starts to become brighter), clamp in position.
7. Cut the hepatic portal vein (if it does not work, try to put the needle into the portal vein).
8. The liver should discolor rapidly becoming a creamy color, if it is very slow to discolor or red patches remain, the liver should not be used.
9. Once the perfusion of the mouse has begun, wash the liver with ~100 mL HBSS-EGTA solution (takes approximately 10 min). Then stop the pump and transfer the tubing to the 50 mL Falcon tube with collagenase buffer.
10. Perfuse the liver until appearance of being well digested (it will look spotty, less quick to rebound to touch, and start to become more fragile after checking with a cotton tip) after approximately 25–30 mL of collagenase buffer.
11. When the digestion is done, stop the perfusion, excise the liver by removing the gall bladder and all connecting ligaments and place it into a petri dish.
12. Add cold HBSS-$CaCl_2$ solution to the liver and release the cells by cutting the liver lobes using the scissors, continue until big clumps are gone but be careful not to cut too much.
13. Filter the cells through the 100 μm cell strainer into a 50 mL Falcon tube. Fill up the tube with cold HBSS-$CaCl_2$ solution to a maximum of 15–20 mL, keep the cell suspensions on ice until all liver samples are ready for the following procedure.

#### *3.1.2 Enrichment of Liver Macrophages by Density Gradient*

Perform the Following Steps in a Cell Culture Hood

1. Centrifuge the liver samples for 3 min at 50 × *g* at 4 °C to collect the hepatocytes. Transfer the supernatant containing the NPCs (approx. 15 mL) into a new 50 mL Falcon tube (be careful not to touch the pellet with hepatocytes).
2. Slowly and carefully add the supernatant on top of the Percoll gradient without disturbing the two layers using a 10 mL serological pipette. Centrifuge the sample at 2300 × *g* for 30 min without brakes.

3. Aspirate the middle interphase with a pipette and transfer into a new 50 mL tube (you can see some cells floating in the middle of the tube, the upper phase is remnants of hepatocytes, on the bottom is debris and red blood cells).
4. Wash two times with 20 mL of 1× PBS and centrifuge at 1500 × *g* for 10 min in RT. Discard the supernatant.
5. After the last wash, resuspend the cell pellet in 2 mL of RPMI media supplemented with 10% FBS, 1% P/S. Count the cells and seed approximately $2 \times 10^6$ cells per well in a 6-well culture plate.
6. Allow the cells to attach for 30 min in a cell culture incubator at 37 °C (5% $CO_2$).

### 3.2 RNA Extraction from Isolated Liver Macrophages

#### 3.2.1 RNA isolation

*The following steps should be performed in a chemical fume hood. All centrifugations are performed at 4 °C, unless otherwise specified.*

1. Start by spraying the working area and equipment with RNase*Zap*.
2. Remove growth media from the culture dish/plate.
3. Add 1 mL of TRIzol reagent per well (for a 6 well plate) to the cells.
4. Lyse the cells by pipetting up and down several times, make sure to get everything off the plate.
5. Transfer the cell lysate to a 1.5 mL Eppendorf tube and add 200 μL of chloroform. Invert the tube up and down 10 times to mix.
6. Centrifuge the sample at 12,000 × *g* for 15 min at 4 °C.
7. Carefully transfer the upper aqueous phase containing the RNA to a fresh tube. Angle the tube at 45° to pipette out the solution.
8. Add 1 μL of GlycoBlue to the aqueous phase.
9. Add an equal volume of isopropanol as to the aqueous phase in a 1:1 ratio.
10. Invert the tube up and down 4 times and incubate at −80 °C overnight.
11. Allow samples to thaw and centrifuge at 12,000 × *g* for 30 min.
12. Remove the supernatant from the tube, leaving the blue RNA pellet.
13. Wash the sample twice by adding 500 μL of 70–80% ethanol and then centrifuge at 7500 × *g* for 10 min.
14. Carefully discard the wash and remove the rest with a P20 pipette.

15. Air-dry the RNA pellet for 5–10 min. Do not allow the RNA to dry completely, because the pellet can lose solubility. Partially dissolved RNA samples have an $A_{260/280}$ ratio < 1.6.
16. Resuspend the pellet in 17 μL of nuclease-free water by pipetting up and down (*see* **Note 2**).
17. Immediately place the tubes on ice! Samples are stored at −80°C.

#### 3.2.2 Quantification and Quality Assessment of RNA Samples

1. Thaw RNA samples on ice. Gently flick the bottom of the tubes to homogenize the RNA sample (*see* **Note 3**) and collect the sample by brief centrifugation in a benchtop minicentrifuge.
2. Measure the RNA concentration using the Qubit High-Sensitivity RNA assay according to the manufacturer's instructions.
3. Run a Bioanalyzer RNA Nano 6000 chip according to the manufacturer's instructions and run using 1 μL of the RNA sample. Select RNA samples with RIN values ≥8 and sufficient RNA content (0.1–1 μg) for use in bulk library preparation.

### 3.3 Library Preparation Using the Illumina TruSeq Stranded mRNA LT Sample Prep Kit (See Illumina Protocol for Further Details) and Data Analysis

The protocol below is adapted from the Illumina TruSeq Stranded mRNA library preparation protocol, with minor modifications and describes the procedures for low number of samples (less than 24 samples).

The entire library preparation extends over 2 days. Unless a safe stopping point is specified in the protocol, the next step should be carried out immediately (*see* **Note 4**).

#### 3.3.1 Poly-A RNA Purification

*For efficiency use a multichannel pipette to mix and aspirate solutions. Always preheat the thermal cycler lid before use. All centrifugations in this library preparation occur between 15 and 25 °C, unless stated otherwise.*

1. Determine the input amount of total RNA (0.1–1 μg) to be used for the library preparation (*see* **Note 5**).
2. Thaw all reagents needed for the RNA purification for 30 min at RT.
3. In a hard-shell 96-well PCR plate, dilute the total RNA with nuclease-free water to 50 μL. Keep the plate on ice, until **step 5**.
4. Vortex the RNA purification beads (oligo-dT magnetic beads) vigorously until the solution is homogeneous.
5. Add 50 μL of beads to each well containing an RNA sample. Gently mix by pipetting up and down six times, until homogeneous.

6. Seal the plate with an adhesive plastic PCR seal.
7. Insert the plate in the thermal cycler and start the "mRNA Denaturation program" (*see* Table 1) to denature the mRNA and allowing the oligo dT beads to bind the polyadenylated RNA molecules.
8. Proceed by incubating the plate at RT for 5 min.
9. Centrifuge the plate at 280 × *g* for 1 min (*see* **Note 6**).
10. Place the plate on 96-well magnetic stand and incubate for 5 min at RT, or until the solution turns clear. Then carefully aspirate and discard the solution from each well without disturbing the pelleted beads. Remove the plate from the magnetic stand and add 200 μL of Bead Washing Buffer (BWB) to each well. Resuspend the beads by gently pipetting up and down six times.
11. Place the PCR plate on the magnetic stand again and incubate for 5 min. Completely aspirate the supernatant from each well.
12. Remove the PCR plate from the magnetic stand and resuspend the beads in 50 μL of Elution Buffer (ELB). Mix thoroughly by gently pipetting up and down ten times, until homogeneous.
13. Centrifuge at 280 × *g* for 1 min.
14. Seal the plate and run the "mRNA Elution 1 program" (*see* Table 2) to detach the isolated poly-A mRNA from the beads.
15. Once run has completed, add 50 μL of Bead Binding Buffer (BBB) to each well; allowing mRNA to bind to the beads again to increase the amount of specific binding of mRNA and to reduce nonspecific binding of contaminating rRNA. Gently mix by pipetting up and down six times.
16. Incubate the plate at RT for 5 min.
17. Then incubate the plate on the magnetic stand for 5 min. Aspirate and discard the supernatant from each well.
18. Remove the plate from the magnetic stand.
19. Add 200 μL of BWB to each well. Resuspend the beads by gently pipetting up and down 20 times, or until homogeneous. Incubate the plate on the magnetic stand for 5 min and aspirate the supernatant.
20. Remove the plate from the magnetic stand and resuspend the beads in 19.5 μL of Fragment, Prime, Finish Mix (FPF) containing the first-strand synthesis reaction buffer by gently pipetting up and down 6 times.
21. Seal the plate and place inside the thermal cycler.
22. Run the "Elution 2-Frag-Prime program" (*see* Table 3) to elute, fragment and prime the mRNA.
23. Centrifuge briefly to collect the condensation.

*3.3.2 First-Strand cDNA Synthesis*

24. Thaw the First Strand Synthesis Act D Mix (FSA) at RT.
25. Incubate the PCR plate on the magnetic stand for 5 min at RT, until the solution turns clear.
26. Keep the plate on the magnetic stand and transfer 17 μL of the supernatant from each well to a new PCR plate.
27. Briefly centrifuge the FSA using a benchtop minicentrifuge.
28. Mix SuperScript II RT and FSA at a ratio of 1:10 (1 μL SuperScript II RT for each 9 μL of FSA), and mix gently (8 μL of final solution will be used per well). Prepare mix for the exact number of samples.
29. Briefly centrifuge the prepared FSA and SuperScript II RT mix and add 8 μL per well. Mix gently by pipetting up and down six times.
30. Seal the PCR plate and collect the content of the wells by centrifuging at 280 × *g* for 1 min.
31. Insert the plate in the thermal cycler and run the "Synthesize 1$^{st}$ Strand program" (*see* Table 4).

*3.3.3 Second-Strand cDNA Synthesis*

32. Thaw Second Strand Marking Master Mix (SMM) at RT. Return to −20 °C after use.
33. Equilibrate AMPure XP beads at RT for at least 30 min.
34. Add either 5 μL of diluted (1:50 in Resuspension buffer) End Repair Control (in-line control reagent) or 5 μL of Resuspension buffer (RSB) to each well. The End Repair Control (CTE) is optional (*see* **Note 7**) and can be omitted; replace with equal volume of RSB instead.
35. Briefly centrifuge the SMM using a benchtop minicentrifuge.
36. Add 20 μL of SMM to each well and gently mix by pipetting up and down 6 times.
37. Seal the plate and briefly centrifuge at 280 × *g* for 10–15 s.
38. Insert the plate inside the thermal cycler and run the "16 °C incubation program" (*see* Table 5) for the second-strand synthesis.
39. Place the PCR plate on the bench and allow the plate to cool and reach room temperature.
40. Vortex the AMPure XP beads vigorously and add 90 μL of beads to each well to clean the double-stranded (ds) cDNA. Gently mix by pipetting up and down ten times.
41. Incubate the plate for 15 min at RT.
42. Centrifuge the plate at 280 × *g* for 1 min.
43. Place the plate on the magnetic stand and incubate for 5 min, until the liquid turns clear.

44. Remove 135 μL of the supernatant. A total of 140 μL of solution is in each well.
45. Leave the plate on the magnetic stand and wash the beads by adding 200 μL of freshly prepared 80% EtOH to each well. Incubate for 30 s. Aspirate the supernatant.
46. Repeat the wash with EtOH. Completely aspirate the supernatant and remove residual EtOH using a P20 pipette.
47. After the second wash, allow the beads to dry for a maximum of 15 min at RT. Do not overdry the beads (*see* **Note 8**)!
48. Remove the plate from the magnetic stand and resuspend the beads in 17.5 μL of RSB per well (*see* **Note 9**). Mix gently by pipetting up and down ten times. Incubate the plate at RT for 2 min.
49. Incubate the plate on the magnetic stand for 5 min. Then transfer 15 μL of the supernatant containing the double-stranded cDNA to a new PCR plate. **OPTIONAL: SAFE STOPPING POINT** (seal and store at −20 °C).

#### *3.3.4 Adenylate 3′-Ends*

1. Thaw A-Tailing Mix (ATL) at RT. Return to −20 °C after use.
2. Add 2.5 μL of the diluted A-Tailing Control (CTA; 1:100 in RSB) to each well if using the in-line control reagent (CTE), otherwise replace the diluted CTA with 2.5 μL of RSB (*see* **Note** 7).
3. Briefly centrifuge the ATL using a benchtop minicentrifuge.
4. Then add 12.5 μL of ATL to each well and mix by gently pipetting up and down 10 times.
5. Seal the plate and centrifuge at 280 × *g* for 1 min.
6. Insert the plate in the thermal cycler and run the "ATAIL70 program" (*see* Table 6).
7. Then centrifuge the plate at 280 × *g* for 1 min.

#### *3.3.5 Ligating the Index Adapters to the ds-cDNA*

Equilibrate the AMPure XP Beads for 30 min at RT and Vortex Thoroughly Before Use

1. Select the RNA Adapter Indexes to be used according to the number of samples to be sequenced and pooling strategy.
2. Thaw the RNA Adapter tubes at RT for 10 min.
3. Thaw the Stop Ligation buffer (STL) at RT.
4. Briefly centrifuge the RNA Adapter tubes using a benchtop minicentrifuge.
5. Add 2.5 μL of the diluted Ligation Control (CTL; 1:100 in RSB) or 2.5 μL of RSB to each well. The CTL is optional, replace with equal volume of RSB if the CTL is omitted.
6. Add 2.5 μL of Ligation Mix (LIG) to each well. Return the LIG to −20 °C directly after use.

7. Add 2.5 μL of designated RNA Adapter Index to each well and mix by gently pipetting up and down 10 times. Return the RNA Adapter Indexes to −20 °C after use.
8. Seal the plate and centrifuge at 280 × *g* for 1 min to collect the content.
9. Insert the plate in the thermal cycler and run the "LIG program" (*see* Table 7).
10. Remove the adhesive seal from the plate and add 5 μL of STL to each well and mix by gently pipetting up and down 10 times, in order to inactivate the ligation.
11. Clean the ds-cDNA using AMPure XP beads. Vortex the beads thoroughly for 1–2 min until homogeneous. Then slowly aspirate and dispense 42 μL of beads to each well and mix by gently pipetting up and down 10 times. Incubate the plate at RT for 15 min.
12. Thereafter, place the plate on the magnetic stand for 5 min, until the liquid turns clear. While on the magnetic stand, carefully aspirate and discard the supernatant without disturbing the beads.
13. Keep the plate on the magnetic stand. Add 180 μL of freshly prepared 80% EtOH and incubate for 30 s. Then remove and discard the supernatant without disturbing the beads. Repeat the wash again with fresh EtOH.
14. After the last wash, allow the samples to air-dry for 8–15 min. Do not overdry the beads (*see* **Note 8**)!
15. Then remove the plate from the magnetic stand and add 52.5 μL of RSB to each well and mix by gently pipetting up and down until the beads have been fully resuspended.
16. Incubate the plate for 2 min at RT and then place it on the magnetic stand again for another 5 min until the liquid turns clear. Collect 50 μL of the supernatant and transfer to a new PCR plate.
17. Vortex the AMPure XP Beads again and add 50 μl of beads to the transferred samples, mix by pipetting up and down until solution becomes fully homogeneous. Incubate the plate for 15 min. Then place the plate on the magnetic stand and incubate for 5 min. Aspirate and discard the supernatant.
18. While on the magnetic stand, add 180 μL of 80% EtOH and incubate for 30 s. Then remove and discard the supernatant without disturbing the beads. Repeat the wash with fresh EtOH.
19. After the last wash, allow the samples to air-dry for 8–15 min. Then remove the plate from the magnetic stand and add 22.5 μL of RSB to each well and mix by gently pipetting up

and down 10 times until the beads have been fully resuspended.

20. Incubate the plate for 2 min at RT and then place it on the magnetic stand again for another 5 min until the liquid is clear. Collect 20 μL of the supernatant and transfer to a new PCR plate.
21. Gently tap the plate on the bench to collect the content of the wells. Seal the plate with an adhesive PCR seal. **OPTIONAL: SAFE STOPPING POINT (store at −20 °C).**

#### *3.3.6 Enrichment PCR to Amplify the Indexed cDNA Libraries*

Equilibrate the AMPure XP Beads for 30 min at RT and Vortex Thoroughly Before Use

1. Thaw the PCR Primer Cocktail (PPC) at RT.
2. Thaw the PCR Master Mix (PMM) on ice.
3. Gently invert the PPC and PMM tubes to mix and briefly centrifuge using a benchtop minicentrifuge. Do not vortex these reagents. Return reagents to −20 °C after use.
4. Place the PCR plate on ice and add 5 μL of PPC to each well. Then add 25 μL of PMM. Mix gently by pipetting up and down ten times.
5. Seal the plate and centrifuge at 280 × *g* for 1 min.
6. Insert the plate in the thermal cycler and run the "Enrichment PCR program" (*see* Table 8).
7. Centrifuge the plate once again at 280 × *g* for 1 min.
8. Thoroughly vortex the AMPure XP beads and dispense 50 μL of beads to each well. Mix gently by pipetting up and down ten times, or until solution becomes completely homogeneous. Incubate the plate for 15 min at RT. Then place the plate on the magnetic stand for 5 min, until the solution turns clear.
9. While on the magnetic stand, aspirate and discard the supernatant of each well. Add 180 μL of freshly prepared 80% EtOH to each well and incubate for 30 s. Then remove and discard the supernatant. Repeat the wash with EtOH. Make sure to aspirate any residual volume of EtOH.
10. After the last wash, keep the plate on the magnetic stand and allow the samples to air-dry for 8–15 min. Then remove the plate from the magnetic stand (*see* **Note 10**) and resuspend the beads in 32.5 μL of RSB (*see* **Note 11**) by pipetting up and down 10 times.
11. Incubate the plate for 2 min at RT. Then, place it back on the magnetic stand and incubate for 2–5 min, until the solution turns clear.
12. Transfer 30 μL of the supernatant to a new PCR plate. **SAFE STOPPING POINT** (seal and store at −20 °C for up to 7 days, *see* **Note 12**).

#### 3.3.7 Quality Control, Normalization, and Pooling of cDNA Libraries

1. Thaw the library plate on ice and briefly centrifuge at 4 °C.
2. Assess the quality of the libraries using the Agilent Technologies 2100 Bioanalyzer with the High-Sensitivity DNA chip (*see* **Note 13**). Add 1 μL of library to the chip. Evaluate the size and purity of your samples, the size of the libraries should be around 260 bp.
3. Quantify the concentration of your libraries with qPCR using the Illumina Sequencing Library qPCR Quantification Kit, following the manufacturer's instructions (*see* **Note 14**).
4. In the 2100 Expert software, identify the average fragment size of libraries by double-clicking on the library sample of choice. Select the Regions Table and limit the measurements from 200 to 500 bp. The average size will be listed in the table below the electropherogram trace.
5. Convert the library concentration from ng/μL to nanomolar (nM) by entering the library concentration in ng/μL (from the qPCR library quantification) and the average fragment size (from the Bioanalyzer) with a nM converter (Google search: "nM conversion calculator excel" to obtain an easy and ready-to-use Excel spreadsheet) or calculate by hand following the instructions from the Illumina quantification kit.
6. Dilute the library samples to 10 nM (based on the calculations from **step 5**) with the Resuspension Buffer (RSB) provided from the library preparation kit.
7. In 0.2 mL PCR tubes or 1.5 mL Eppendorf tubes, multiplex libraries by pooling an equal volume of each 10 nM library based on your pooling strategy (*see* **Note 15**).
8. Further dilute the library pools to the required concentration prior to the sequencing, using the RSB.
9. Sequence samples with an Illumina sequencer.

### 3.4 Bioinformatic Analyses of Next Generation Sequencing Data

*Analyses of RNA sequencing data utilize the free, open-source R software and Bioconductor packages. Bioinformatic analysis pipelines for bulk RNA sequencing aligns, normalizes, and counts sequenced reads. The pipeline also generates visualization plots and performs differential gene expression analysis in order to identify genes of interest.*

1. In brief, the sequencing data is first demultiplexed using the known index adapter sequences that was ligated to the cDNA fragments.
2. Then, the FASTQ files (containing the nucleotide sequence of each read with a quality score) are aligned to a reference genome.
3. Each read is then counted, normalized and summarized as an expression matrix.

4. The expression matrix is used for visualization by graph plotting and differential gene expression analysis is performed.
5. The generated results (a table of genes, their expression value and statistical significance) can then be exported as a CSV file and opened as an Excel spreadsheet.

## 4 Notes

1. Selecting between sets of (Set A or Set B) index adapters depends on the multiplexing (or pooling) strategy of libraries. For 2-plexing of libraries, the manufacturer does not recommend the Set B indexes; opt here for the Set A index adapters instead.
2. The RNA pellet can sometimes be difficult to resuspend. After adding the water, incubating the RNA samples on ice for 1 h or longer facilitates to dissolve the RNA pellet. Alternatively, incubate RNA samples at 65 °C for 10 min or freeze samples at −80 °C and thaw again.
3. Avoid vortexing the RNA sample!
4. To ensure that all procedures will run smoothly, we recommend that all the required thermal cycler programs (*see* Tables 1, 2, 3, 4, 5, 6, 7, and 8) have been preinstalled and saved in the thermal cycler prior to generating the libraries. For easier navigation, save the programs in a designated master folder in the thermal cycler.
5. Anything within the given range of RNA (0.1–1 μg) can be used for this library preparation kit. We have used as low as 150 ng of macrophage RNA for the bulk library preparation. However, to ensure optimal outcome, it is better to avoid using 1 μg of RNA (anything lower than this is more suitable) as discrepancies in measuring and pipetting errors can result in underfragmented RNA and final cDNA libraries of larger than expected size distribution.
6. Centrifuging the plate at 4 °C results in a slight sedimentation of the RNA-purification beads. Make sure that the centrifuge is set between 15 and 25 °C.
7. The controls can only be used if the analysis pipeline used has been designed to identify these. In our case, we excluded all the controls and replaced the same volume with RSB instead.
8. Do not let the beads overdry! The beads are prone to breakage and detach from the wells with any sudden movement involving the slightest air flow. In our experience, the beads start to crack already after 7.5–8 min of drying. The optimal drying time is once small cracks in the beads are present.

9. The beads can be difficult to resuspend. Therefore, add the RSB directly on to the dried beads, in alternating order of wells, that is, wells #1, 3, 5, 7, and so on. Then, rotate the plate 180° and add the RSB to wells #2, 4, 6, 8, and so on. Allow the moisture from the RSB to bring and collect the beads to the bottom of the wells. Proceed by resuspending the beads by pipetting.
10. When removing the plate from the magnetic stand, the air-dried beads will crumble. Do not make any sudden movements to the plate that can cause the beads to detach. The beads and thus the samples could end up lost or inside another well!
11. Add the RSB directly on the dried bead and incubate for 30–60 s at RT. This will ensure easier resuspension of the beads.
12. Although the original Illumina protocol states that the plate can be stored for up to 7 days, we have kept the plate for longer prior to the sequencing and still obtained good results.
13. Avoid generating air bubbles when running a Bioanalyzer chip. Air bubbles present in the gel can inhibit the gel electrophoresis. If vortexing does not help, the air bubbles can sometimes be burst by using an empty 1000 μL pipette tip attached to a P1000 pipette set at around 500 μL. Gently burst the bubble by repeatedly pipetting above the affected well(s). Alternatively, completely aspirate the gel from the well and load with new gel.
14. Library quantification with this qPCR assay is the most accurate measurement as it detects only double-stranded cDNA fragments with index adapters. Only indexed libraries will generate successful sequencing reads. To calculate the library concentrations with the qPCR quantification, the average fragment size/molecule length of the libraries is required. This is obtained from the Bioanalyzer data of the libraries.
15. To minimize index hopping, do not store libraries as pooled. Pool libraries only prior to sequencing.

**Table 1**
**mRNA denaturation thermal cycler program**

| **mRNA Denaturation program** | | |
|---|---|---|
| **Lid temperature: 100 °C** | | |
| **Step** | **Temperature** | **Duration** |
| 1 | 65 °C | 5 min |
| 2 | 4 °C | ∞ |

**Table 2**
**mRNA elution thermal cycler program**

| mRNA Elution 1 program | | |
|---|---|---|
| Lid temperature: 100 °C | | |
| Step | Temperature | Duration |
| 1 | 80 °C | 2 min |
| 2 | 25 °C | ∞ |

**Table 3**
**Thermal cycler program for the fragmentation of mRNA**

| Elution 2-Frag-Prime program | | |
|---|---|---|
| Lid temperature: 100 °C | | |
| Step | Temperature | Duration |
| 1 | 94 °C | 8 min |
| 2 | 4 °C | ∞ |

**Table 4**
**First-strand cDNA synthesis thermal cycler program**

| Synthesize 1st Strand program | | |
|---|---|---|
| Lid temperature: 100 °C | | |
| Step | Temperature | Duration |
| 1 | 25 °C | 10 min |
| 2 | 42 °C | 15 min |
| 3 | 70 °C | 15 min |
| 4 | 4 °C | ∞ |

**Table 5**
**16 °C incubation thermal cycler program for second-strand cDNA synthesis**

| 16 °C incubation program | | |
|---|---|---|
| Lid temperature: 30 °C | | |
| Step | Temperature | Duration |
| 1 | 16 °C | 60 min |

**Table 6**
**3′-ended adenylation thermal cycler program**

| ATAIL70 program | | |
|---|---|---|
| Lid temperature: 100 °C | | |
| **Step** | **Temperature** | **Duration** |
| 1 | 37 °C | 30 min |
| 2 | 70 °C | 5 min |
| 3 | 4 °C | ∞ |

**Table 7**
**Index adapter ligation thermal cycler program**

| LIG program | | |
|---|---|---|
| Lid temperature: 100 °C | | |
| **Step** | **Temperature** | **Duration** |
| 1 | 30 °C | 10 min |
| 2 | 4 °C | ∞ |

**Table 8**
**Enrichment PCR thermal cycler program for amplification of indexed cDNA fragments**

| Enrichment PCR program | | |
|---|---|---|
| Lid temperature: 100 °C | | |
| **Step** | **Temperature** | **Duration** |
| 1 | 98 °C | 30 s |
| 2 | 98 °C | 10 s |
| | 60 °C | 30 s |
| | 72 °C | 30 s |
| 3 | Repeat **step 2**, with 14 additional cycles | |
| 4 | 72 °C | 5 min |
| 5 | 4 °C | ∞ |

## References


1. Crispe IN (2011) Liver antigen-presenting cells. J Hepatol 54(2):357–365. https://doi.org/10.1016/j.jhep.2010.10.005
2. Racanelli V, Rehermann B (2006) The liver as an immunological organ. Hepatology 43(2 Suppl 1):S54–S62. https://doi.org/10.1002/hep.21060
3. Thomson AW, Knolle PA (2010) Antigen-presenting cell function in the tolerogenic liver environment. Nat Rev Immunol 10(11):753–766. https://doi.org/10.1038/nri2858
4. Mowat AM, Scott CL, Bain CC (2017) Barrier-tissue macrophages: functional adaptation to environmental challenges. Nat Med 23(11):1258–1270. https://doi.org/10.1038/nm.4430
5. Andrew C, Brian DB, Miriam M (2011) Studying the mononuclear phagocyte system in the molecular age. Nat Rev Immunol 11(11):788. https://doi.org/10.1038/nri3087
6. Hume DA (2011) Applications of myeloid-specific promoters in transgenic mice support in vivo imaging and functional genomics but do not support the concept of distinct macrophage and dendritic cell lineages or roles in immunity. J Leukoc Biol 89(4):525–538. https://doi.org/10.1189/jlb.0810472
7. Bassler K, Schulte-Schrepping J, Warnat-Herresthal S, Aschenbrenner AC, Schultze JL (2019) The myeloid cell compartmentcell by cell. Annu Rev Immunol 37(1):269–293. https://doi.org/10.1146/annurev-immunol-042718-041728

# Chapter 6

# Kupffer Cell and Monocyte-Derived Macrophage Identification by Immunofluorescence on Formalin-Fixed, Paraffin-Embedded (FFPE) Mouse Liver Sections

Adrien Guillot, Chloé Buch, and Tony Jourdan

## Abstract

Kupffer cells are the liver-resident macrophages and represent the first line of defense between the pathogens circulating from the intestines through the portal vein and systemic circulation. Recent works have highlighted the complex heterogeneity of macrophage functions and origins, thus raising awareness on the need for a better characterization of macrophage populations. The immunohistochemistry method here described, allows for a rapid distinction between Kupffer cells and monocyte-derived macrophages present on formalin-fixed, paraffin-embedded mouse liver samples. This protocol has been optimized for its reproducibility, reliability, and simplicity.

**Key words** Kupffer cell, Liver, Monocyte, Macrophage, Immunofluorescence, CLEC4F/CLECSF13, IBA1, Confocal microscopy, FFPE liver sections

## 1 Introduction

The liver is a vital organ with crucial metabolic and immune functions [1, 2]. Hepatocytes represent the predominant cell type of the liver. They are mostly involved in the control of lipid, carbohydrate, and protein metabolism, as well as the elimination of toxic compounds through enzymatic activity and bile secretion [3–5]. Liver macrophages are represented by tissue-resident macrophages or Kupffer cells (KCs) and recruited monocyte-derived macrophages (MoMs). They have been implicated in the regulation of multiple aspects of liver biology, ranging from maintaining an immunotolerant environment to aggravating systemic inflammation, favoring liver regeneration and increasing fibrosis [6, 7]. KCs represent about 20% of the nonparenchymal cell population in mouse liver and are located in the lumen of liver sinusoids. They play an important role in maintaining a tolerogenic environment in the healthy liver and are essential for the phagocytosis of circulating

Myriam Aouadi and Valerio Azzimato (eds.), *Kupffer Cells: Methods and Protocols*, Methods in Molecular Biology, vol. 2164, https://doi.org/10.1007/978-1-0716-0704-6_6, © Springer Science+Business Media, LLC, part of Springer Nature 2020

pathogens and cellular debris. During liver injury, KCs are activated and secrete various cytokines and chemokines, thus inducing the recruitment of circulating immune cells such as monocyte-derived macrophages (MoMs) within the liver [6]. However, the distinction between the KCs and MoMs remains challenging. Here, we describe a simple immunostaining protocol allowing to distinguish KCs from MoMs in mouse liver formalin-fixed, paraffin-embedded (FFPE) tissue sections (*see* **Note 1**). This method is based on dual fluorescent labeling of the calcium-binding protein IBA1 specifically expressed in macrophage/microglia and regarded as a panmacrophage marker [8, 9], and the C-type lectin domain family 4 member F (CLEC4F/CLECSF13, termed as CLEC4F below) is used as a specific KC marker [10, 11].

## 2 Materials

### 2.1 Mouse Euthanasia and Tissue Collection

1. Pentobarbital sodium, phosphate buffer solution (PBS).
2. Dissection forceps and scissors.
3. 70% ethanol.
4. 1× PBS buffer: PBS-10×, distilled water.

### 2.2 Liver Fixation, Tissue Dehydration, and Paraffin Embedding

1. Formalin, buffered, 10% (Phosphate Buffer/Certified).
2. Xylene.
3. Ethanol pure grade.
4. 50% xylene–ethanol: 500 mL xylene, 500 mL ethanol pure grade.
5. Ethanol 95%: 950 mL ethanol pure grade, 50 mL ultrapure water.
6. Ethanol 80%: 800 mL ethanol pure grade, 200 mL ultrapure water.
7. Ethanol 70%: 700 mL ethanol pure grade, 300 mL ultrapure water.
8. TissuePrep™ 2 (ThermoFisher #T555).
9. Plus Charged Microscope Slides or equivalent.
10. Tissue Embedding and Processing Cassettes Stainless Steel Base Mold.
11. Thermo Scientific™ Tissue Path™ IV Tissue Cassettes.

### 2.3 Immunostaining Reagents

1. 1× PBS buffer: PBS-10×, distilled water.
2. 1× citrate buffer: 10× citrate buffer, distilled water.
3. 3% normal goat serum (NGS-PBS): Normal goat serum, 1× PBS buffer.

4. Primary antibodies: Prepare extemporary by mixing 5 μg/mL rat anti-CLEC4F/CLECSF13 antibody (R&D #MAB2784, stock solution reconstituted at 0.5 mg/mL in sterile 1× PBS, aliquoted and stored at −20 °C) and 1 μg/mL rabbit anti-IBA1 antibody (VWR #100369-764) in 3% NGS-PBS (*see* **Note 2**).
5. Secondary antibodies: Prepare extemporary by mixing 4 μg/mL anti-rabbit IgG (H + L) (Alexa Fluor® 488 Conjugate, Cell Signaling #4412) and 4 μg/mL anti-rat IgG (H + L), (Alexa Fluor® 555 Conjugate, Cell Signaling #4417) in 3% NGS-PBS (*see* **Notes 3** and **4**).
6. Nucleic acid staining (nucleus): DAPI: prepare a 20 mg/mL stock solution in water, then dilute it 1.1000 to generate a 20 μg/mL working solution in 1× PBS.
7. Mounting medium: VectaMount AQ Aqueous Mounting Medium.

## 3 Methods

### 3.1 Mouse Euthanasia, Tissue Collection, and Fixation

1. The mouse is euthanized with a lethal dose of pentobarbital sodium (90 mg/kg, intraperitoneally) (*see* **Note 5**).
2. Once the mouse no longer responds to toe pinch, perform an exsanguination by the preferred technique (e.g., traumatic avulsion of the orbital globe), followed by cervical dislocation.
3. Place the animal on its back on a dissection board, wipe the abdomen with ethanol, and perform a laparotomy to expose the liver.
4. Dissect the liver out, briefly rinse in 1× PBS and collect one specimen of each lobe for histology. Each specimen should be about 25 $mm^3$ (*see* **Note 6**).
5. To preserve tissue morphology and retain the antigenicity of the target molecules, incubate samples in 3 mL formalin (10%) for 24–36 h at room temperature, under gentle agitation (*see* **Note 7**).

### 3.2 Tissue Embedding and Processing

1. Collect your samples and place them in a histology cassette labeled with the sample ID.
2. All the cassettes containing samples should be immerged in three consecutive tap water baths for 5 min, in order to rinse off the formalin.
3. Place the cassettes in a bath of ethanol 70% and follow the sequence described in Table 1.

**Table 1**
**Dehydration and embedding of the specimens**

| Step | Solution | Incubation time (h) |
|---|---|---|
| 1 | Ethanol 70% | 1 |
| 2 | Ethanol 80% | 1 |
| 3 | Ethanol 95% | 1 |
| 4 | Ethanol pure grade | 1 |
| 5 | Ethanol pure grade | 1 |
| 6 | Xylene | 1 |
| 7 | Xylene | 1 |
| 8 | Melted paraffin | 1 |
| 9 | Melted paraffin | 1 |

4. Place the specimen in the center of the mold and embed it in paraffin at 58 °C. Add a printed label on top and let it solidify on a cooling bench (Fig. 1) (*see* **Note 8**).
5. FFPE liver sections are cut at a 4-μm thickness using a microtome (Fig. 2a).
6. Float the sections on a water bath at 56 °C.
7. Mount the sections on the microscope slides.
8. Dry the slides overnight in a ventilated incubator set at 37 °C (*see* **Note 8**).

### 3.3 Immunostaining

#### 3.3.1 Deparaffinization, Rehydration of the Sections, and Antigen Retrieval

1. To perform this critical step, each section will undergo three consecutive 5 min incubations in xylene, followed by incubation in several baths of decreasing ethanol concentrations as described in Table 2 and shown in Fig. 2b. Once the slides have been washed in the above sequence, place slides in distilled water to rinse off ethanol (*see* **Note 9**).
2. Immerse the slides in 1× Citrate buffer, in a small plastic slide holder (Fig. 2c) (*see* **Note 10**).
3. Heat at maximum power using a 1200 W microwave until boiling (approx. 1 min).
4. Let the slides cool down for another minute.
5. Heat again at the lower power, for 10 min.
6. Let the slides to cool down in the citrate buffer at room temperature for 35 min, covered (*see* **Note 11**).

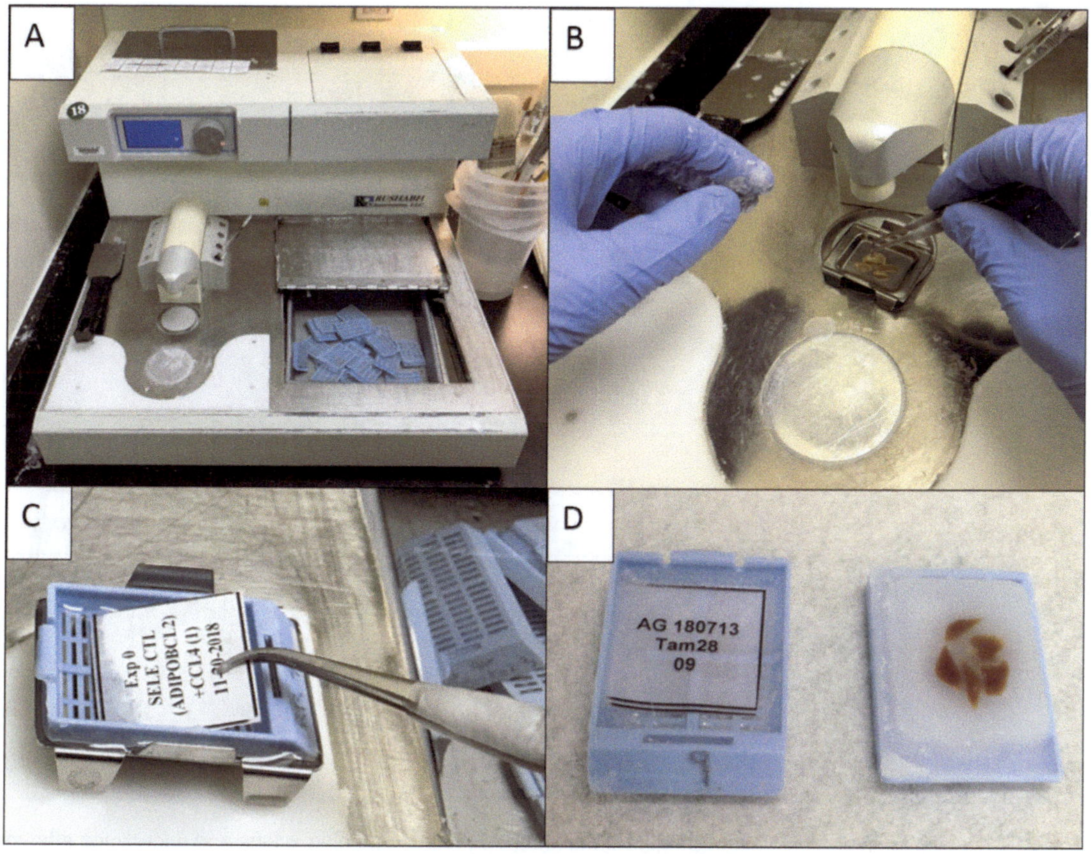


**Fig. 1** Tissue-embedding setup. (**a**) Paraffin embedding machine. After dehydration, samples are soaked in a liquid paraffin bath (heated at 58 °C) and cassette paper labels are prepared. (**b**) Samples are picked up from the plastic cassette and then soaked in liquid paraffin and arranged in the middle of the mold. (**c**) Before paraffin polymerization, add paper label to the cassette to ensure future sample identification. (**d**) Labeled cassette containing liver samples arranged in the middle of the block

*3.3.2 Nonspecific Antigen Blocking and Primary Antibody Incubation*

1. Rinse the slides in 1× PBS, 3× 2 min under gentle orbital agitation.
2. Circle your specimens with a hydrophobic PAP pen and add the 3% NGS-PBS on top of your specimens (*see* **Note 12**).
3. Incubate for 1 h at room temperature.
4. Remove the extra blocking buffer by inverting the slides carefully.
5. Without rinsing the slides, add the primary antibody mixture containing the rabbit anti-IbBA1 and the rat anti-CLEC4F on top of your specimens (*see* **Note 13**).
6. Incubate overnight at 4 °C in a humidified chamber (*see* **Note 14**).

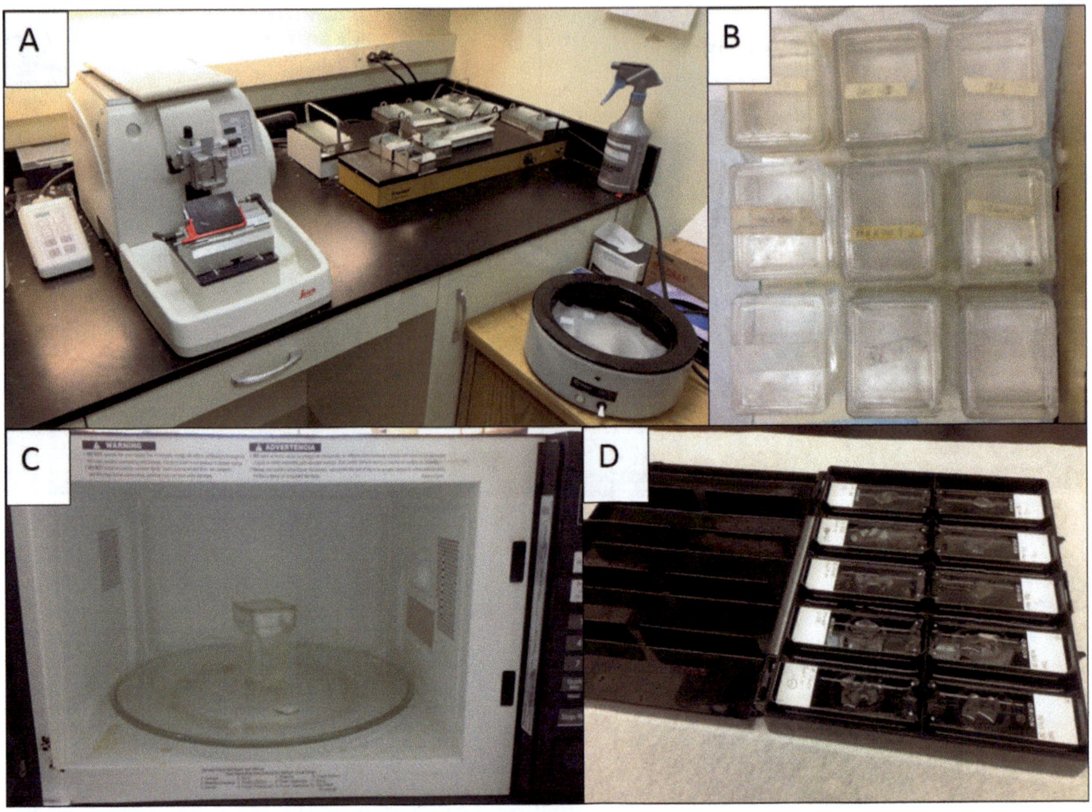


**Fig. 2** Tissue sectioning and staining. (**a**) Microtome setup. (**b**) Jars containing xylene and decreasing ethanol concentration for sample rehydration. (**c**) Antigen unmasking is performed in microwave-heated citrate buffer. (**d**) At all incubation steps, slides are placed in a humidified opaque slide staining tray

**Table 2**
**Deparaffinization and rehydration of the sections**

| Step | Solution | Incubation time |
|---|---|---|
| 1 | Xylene | 5 min |
| 2 | Xylene | 5 min |
| 3 | Xylene | 5 min |
| 4 | 50% Xylene/Ethanol | 20 dips |
| 5 | Ethanol pure grade | 5 min |
| 6 | Ethanol pure grade | 5 min |
| 7 | Ethanol 95% | 5 min |
| 8 | Ethanol 80% | 5 min |
| 9 | Ethanol 70% | 5 min |

*3.3.3 Secondary Antibody Incubation, Nuclear Counterstaining, and Slide Mounting*

1. Rinse the slides in PBS, 3× 2 min under gentle orbital agitation.
2. Apply the secondary antibodies, prepared in 3% NGS-PBS.
3. Incubate for 1 h at room temperature in an opaque humidified chamber (*see* **Note 15**).
4. Rinse the slides in 1× PBS, 3× 2 min.
5. Incubate in DAPI working solution for 10 min at room temperature in an opaque humidified chamber.
6. Rinse the slides in 1× PBS, 3× 2 min.
7. Apply aqueous mounting medium and mount the slides. Staining may be visualized directly, or alternatively let the mounting medium dry overnight at 4 °C, keeping the slides flat.

### 3.4 Confocal Microscopy

We are using a Laser scanning confocal microscope (Zeiss, LSM700) quipped with a fluorescence light source. As illustrated in Fig. 3, all nuclei are stained in blue, IBA1-positive cells appear in green and CLEC4F-positive cells appears red. As a result, the overlay-composite picture will allow us to distinguish yellow KCs (yellow due to the overlay green and red) and green MoMs (*see* **Note 16**).

## 4 Notes

1. The primary and secondary antibodies described here may also be used to stain KCs and MoMs in frozen mouse liver sections.
2. Mix by pipetting; avoid using a vortex for solutions containing antibodies.
3. Any other combination of colors may be used for secondary antibodies but may require additional optimizations.
4. Protect secondary antibodies from light exposure to avoid fluorochrome fading.
5. All animal procedures must be performed by trained personnel and according to the institutional animal welfare guidelines.
6. To avoid mouse-to-mouse discrepancies, always collect the same liver areas in all mice.
7. Agitation will enhance the diffusion of the fixative, thus preventing nonspecific staining in underfixed tissue areas.
8. Paraffin blocks and slides can be stored at room temperature or at 4 °C for several years.
9. Keep the slides in the tap water until ready to perform antigen retrieval. At no time from this point onward should the slides be allowed to dry. Drying out would cause nonspecific antibody binding and therefore high background staining.
10. Never place metallic objects in a microwave.

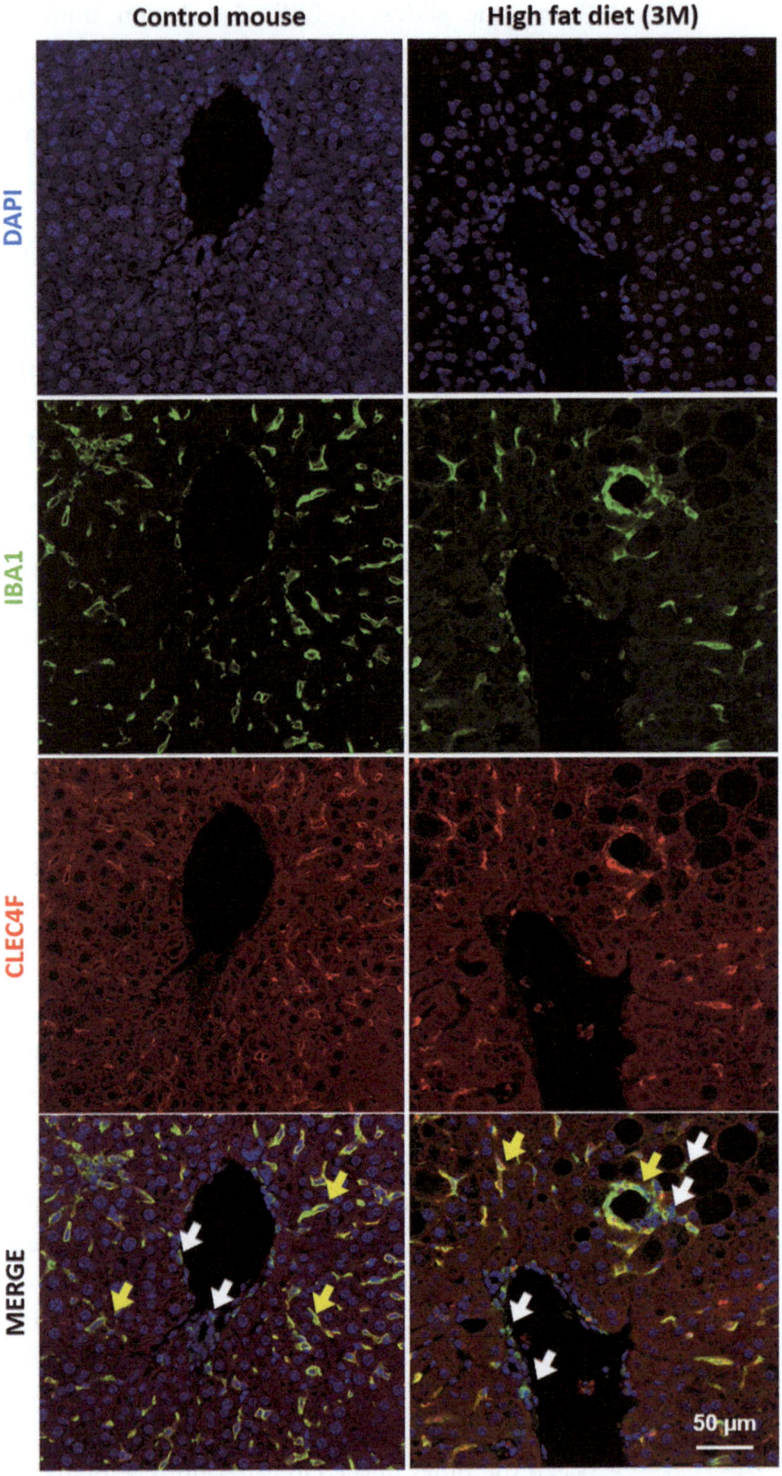


**Fig. 3** Representative CLEC4F and IBA1 immunostaining. FFPE liver sections from control or 3-month high-fat diet–fed mice were stained with anti-CLEC4F and anti-IBA1 antibodies using the method herein described. Kupffer cells appear yellow as a consequence of CLEC4F and IBA1 coexpression (yellow arrows). Circulating and recruited monocytes appear as green (white arrows)

11. The slide container may be hot, use caution when transferring it to the bench.
12. For each slide, 50–200 μL of solutions should be enough to cover the tissue. Adjust volumes depending on sample size.
13. It is critical to include a negative control in which primary antibodies have been omitted to assess nonspecific staining. Such a negative control should be devoid of any signal in the red and green channels.
14. Shorter incubation times may lead to a weaker signal. According to our experience, nonspecific staining is not a concern by using this protocol; thus, overnight incubation is appropriate.
15. From this step, always protect the samples from direct light exposure to prevent photobleaching.
16. In physiologic conditions, CLEC4F-positive cells should also express IBA1 in the liver. Thus, no red cell should be visible on the overlay-composite picture. If so, sample preparation, immunostaining conditions or microscope settings may need to be reassessed. On the other hand, IBA1-expressing CLEC4F-negative (green only) cells may be occasionally observed in liver sinusoids and are usually more likely to be observed in the major blood vessels of central or portal areas.

## References


1. Hotamisligil GS (2006) Inflammation and metabolic disorders. Nature 444:860. https://doi.org/10.1038/nature05485
2. Grunebaum E, Avitzur Y (2018) Liver-associated immune abnormalities. Autoimmun Rev 18(1):15–20. https://doi.org/10.1016/j.autrev.2018.06.016
3. Gissen P, Arias IM (2015) Structural and functional hepatocyte polarity and liver disease. J Hepatol 63(4):1023–1037. https://doi.org/10.1016/j.jhep.2015.06.015
4. Campbell I (2006) Liver: metabolic functions. Anaesth Intensive Care Med 7(2):51–54. https://doi.org/10.1383/anes.2006.7.2.51
5. Malarkey DE, Johnson K, Ryan L, Boorman G, Maronpot RR (2005) New insights into functional aspects of liver morphology. Toxicol Pathol 33(1):27–34. https://doi.org/10.1080/01926230590881826
6. Tacke F (2017) Targeting hepatic macrophages to treat liver diseases. J Hepatol 66(6):1300–1312. https://doi.org/10.1016/j.jhep.2017.02.026
7. Jager J, Aparicio-Vergara M, Aouadi M (2016) Liver innate immune cells and insulin resistance: the multiple facets of Kupffer cells. J Intern Med 280(2):209–220. https://doi.org/10.1111/joim.12483
8. Sasaki Y, Ohsawa K, Kanazawa H, Kohsaka S, Imai Y (2001) Iba1 is an actin-cross-linking protein in macrophages/microglia. Biochem Biophys Res Commun 286(2):292–297. https://doi.org/10.1006/bbrc.2001.5388
9. Rehg JE, Bush D, Ward JM (2012) The utility of immunohistochemistry for the identification of hematopoietic and lymphoid cells in normal tissues and interpretation of proliferative and inflammatory lesions of mice and rats. Toxicol Pathol 40(2):345–374. https://doi.org/10.1177/0192623311430695
10. Tencerova M, Aouadi M, Vangala P, Nicoloro SM, Yawe JC, Cohen JL, Shen Y, Garcia-Menendez L, Pedersen DJ, Gallagher-Dorval-K, Perugini RA, Gupta OT, Czech MP (2015) Activated Kupffer cells inhibit insulin sensitivity in obese mice. FASEB J 29(7):2959–2969. https://doi.org/10.1096/fj.15-270496
11. Yang CY, Chen JB, Tsai TF, Tsai YC, Tsai CY, Liang PH, Hsu TL, Wu CY, Netea MG, Wong CH, Hsieh SL (2013) CLEC4F is an inducible C-type lectin in F4/80-positive cells and is involved in alpha-galactosylceramide presentation in liver. PLoS One 8(6):e65070. https://doi.org/10.1371/journal.pone.0065070

# Chapter 7

# Sample Preparation of Optically Cleared Liver Tissue to Identify Liver Macrophages Using 3D Microscopy

**Otto Strauss and Niklas K. Björkström**

## Abstract

Histology, or the study of tissue microanatomy, is essential to understanding the in situ function of varying cell types within an organ. How cells are distributed throughout organs provides an indication of how they interact with other cells and structures within the organ microanatomy. The tortuous shape and large size of liver macrophages limits the value of standard tissue thickness of 5–10 μm. As a result, imaging of specimens ideally thicker than 100 μm is necessary to investigate the liver microanatomy and the how macrophages are distributed within this. Modern methods of microscopy, such as confocal and light sheet microscopy, allow for the analysis of tissue specimens of a thickness well beyond 100 μm in the *z*-dimension. Liver tissue is an opaque tissue, and as a result, different techniques are needed to ameliorate light diffraction within the tissue. These techniques, in conjunction with antibody staining and refractive index matching of the tissue, have allowed researchers to image liver tissue specimens of more than 100 μm thickness. Two of these techniques are modified versions of the clearing methods known as clearing-enhanced 3D (Ce3D) and fructose, urea, and glycerol for imaging (FUnGI). Here, we discuss the steps involved in preparing tissue specimens for optically clearing tissue using Ce3D and FUnGI for subsequent analysis of the distribution of macrophages in three dimensions using a confocal microscope.

**Key words** Liver, Microscopy, 3D, Clearing, Macrophages

## 1 Introduction

The human liver micro anatomy is a complex structure of repeating units within which different leukocytes, including macrophages, can be identified [1]. Thin sections do not capture entire cells [2] or allow for the appreciation of vascular and biliary structures within the organ. Microscopy of specimens that are large in the *z*-dimensions allows for 3D imaging that provides a framework to truly appreciate where these cells are located in relation to each other and other structures within the organ.

---

**Electronic supplementary material:** The online version of this chapter (https://doi.org/10.1007/978-1-0716-0704-6_7) contains supplementary material, which is available to authorized users.

Myriam Aouadi and Valerio Azzimato (eds.), *Kupffer Cells: Methods and Protocols*, Methods in Molecular Biology, vol. 2164, https://doi.org/10.1007/978-1-0716-0704-6_7, © Springer Science+Business Media, LLC, part of Springer Nature 2020

A major obstacle to overcome in tissue microscopy is the intrinsic optical density of bodily tissue. In recent years methods of tissue clearing have been developed to facilitate whole-mount 3D microscopy. Many types of tissue clearing methods exist [3–7]. In this chapter we present a tissue clearing protocol modified from that described by Li et al. known as "clearing-enhanced 3D"(Ce3D) [8]. We also describe a less toxic method known as fructose, urea, and glycerol for imaging (FUnGI). Once stained and cleared samples can then be imaged using a laser scanning or spinning-disk confocal microscope.

There is a trade-off between maintaining structural aspects of tissue integrity as well as molecular epitopes required for antibodies to target and obtaining optically cleared tissue. We have trialed CLARITY, CUBIC, Ce3D, and FUnGI clearing methods to identify macrophages in human liver. Although CLARITY and CUBIC results in high levels of optical clarity of the tissue [9, 10], they are unfavorable in identifying macrophages as a result of destroying either the tissues structure or molecular epitopes and therefore were not suitable. Ce3D and FUnGI retained both tissue structure and molecular epitopes while providing reduction in tissue opacity.

These methods allow liver macrophages to be identified in three dimensions and thereby obtain a greater appreciation of their locality within the lobule. The liver appears to contain multiple subsets of macrophages [11], how these are distributed throughout the organ has not yet been clearly established. The protocol can be applied in research of other tissues and other cell types, and it has also been trialed on mouse as well as human. Despite this, there are limitations to this method. The most obvious limitation is the availability of appropriate antibodies to identify epitopes of interest. Nonetheless, our method can be used to identify macrophages, other cell types, and structures within the organ, with the potential to be improved as new clearing reagents are developed.

## 2 Materials

### *2.1 Sample Preparation*

1. Histocon tissue transport medium (Histolab).
2. Surgical scalpel.
3. Petri dish, for instance 92 × 16 mm.
4. Paraformaldehyde (PFA) 2%.
5. Phosphate buffered saline (PBS).
6. Razor blades as per Vibratome manufacturer's recommendations (we used double-edge razor blades).

7. Cyanoacrylate adhesive for fixing sample vibratome cutting block.
8. Vibratome: We use a Vibratome 3000; however, there are many commercially available vibratome devices.

### 2.2 Blocking and Staining

1. Blocking buffer: Background Buster (Innovex Biosciences), 0.1% Triton X-100.
2. Staining buffer: tris buffered saline (TBS), 10% v/v human serum, 0.1% v/v Triton X-100.
3. Washing Buffer: TBS, 0.3% v/v Triton X-100.
4. Primary antibodies (as per Table 1).
5. Secondary antibodies (as per Table 2).
6. DAPI.
7. 1.5 mL Eppendorf tubes.

### 2.3 Tissue Clearing media

After blocking and staining steps, samples can be placed in either Ce3D clearing media or FUnGI, the reasons for which are discussed in **Note 1**.

1. Ce3D tissue clearing media (*see* **Note 2**): *N*-methylacetamide 40% (v/v), PBS, Histodenz 86% (w/v), Triton X-100 (0.1% v/v), 1-thioglycerol (0.5% v/v).

**Table 1**
**Primary antibodies used**

| Target | Isotype | Clone | Manufacturer |
|---|---|---|---|
| CD163 | Mouse IgG1 | EDHu-1 | Serotec Ab (now Bio-Rad) |
| Cytokeratin 19 | Rabbit monoclonal | EP1580Y | Abcam |
| LYVE-1 | Rabbit | N/A | Abcam |
| aSMA | Mouse IgG2a | 1A4 | CellMarque |

**Table 2**
**Secondary antibodies used**

| Target | Fluorophore | Species | Laser (nm) |
|---|---|---|---|
| Mouse IgG1 | Alexa Fluor 555 | Goat | 546 |
| Rabbit H + L chain | Alexa Fluor 488 | Goat | 488 |
| Mouse IgG2a | Alexa Fluor 647 | Goat | 641 |

2. FUnGI Clearing media (*see* **Note 1**): 2.5 M fructose, 10 mM Tris, 0.1 mM EDTA, 60% glycerol, 2.5 M Urea, 50 μg/mL ascorbic acid, 0.05 ng/mL L-glutathione.
3. 1.5 mL Eppendorf tubes.
4. Rotating wheel.
5. Parafilm sealant.

#### 2.4 Sample Mounting

1. Silicone grease.
2. #1.5 cover slip (0.16–0.19 mm thickness).
3. Transfer pipette.
4. Glass bottom slides, *see* **Note 3**.
5. Silicone sealant such as picodent twinsil, *see* **Note 3**.

## 3 Methods

All blocking, staining, and clearing can be undertaken in 1.5 mL Eppendorf tubes.

Transfer pipettes are useful for transferring samples at all stages of tissue preparation.

#### 3.1 Fixation

1. Obtain fresh liver sample and place in Histocon buffer. Process immediately or store at 4 °C overnight.
2. Place sample in petri dish, use a surgical scalpel to reduce the size of pieces to approximately 0.5 $cm^3$.
3. Place pieces in falcon tube with 2% PFA at approximately 10× the volume of the samples being fixed and leave over night at 4 °C to fix.
4. After a minimum of 12 h fixation transfer the sample to PBS. Store at 4 °C until further processing.

#### 3.2 Sectioning

1. Use a tissue sectioning device such as a vibratome to reduce thickness of sections to 200 μm thickness (*see* **Note 2**).

#### 3.3 Blocking and Staining

1. Incubate samples in Blocking Buffer overnight on shaker at 37 °C.
2. Add primary antibodies to Staining Buffer at a concentration of 1% v/v, incubate on shaker at 37 °C for between 12 and 36 h. Usually 300 μL of primary staining buffer is required per sample.
3. Replace media with 1 mL of Washing Buffer twice for each sample and then incubate in 1 mL Washing Buffer for 12 h on shaker at 37 °C.

4. Add DAPI and secondary antibodies specific to primary antibodies to Staining Buffer at a concentration of 1% v/v, incubate overnight on shaker at 37 °C. Usually 300 μL of primary staining buffer is required per sample.
5. Replace media with 1 mL of Washing Buffer twice for each sample and then incubate in 1 mL Washing Buffer overnight on shaker at 37 °C.
6. *See* **Note 4** for further information.

### 3.4 Clearing

1. Place sample in new Eppendorf tube and cover with clearing media (usually a minimum of 300 μL). Either Ce3D or FUnGI could be used.
2. Seal tube with Parafilm and incubate on rotating spinning wheel at room temperature for a minimum of 24 h.

### 3.5 Mounting (See Note 3)

1. Within each chamber of a glass bottom well, use the silicone grease to draw a circular well approximately 1 cm in diameter.
2. Use transfer pipette, to place sample in the well, ensure that the sample sits in enough clearing media to fill the well when a coverslip is lowered onto it.
3. Gently lower coverslip onto well, lowering one edge first then the remainder of the coverslip.
4. Apply gentle pressure to coverslip to close the seal with the silicone grease.

### 3.6 Imaging

A detailed description of imaging techniques is beyond the scope of this chapter, imaging can be undertaken on a spinning disk or laser scanning confocal microscope. Rendering of 3D images can be undertaken using a variety of software clients such as Imaris (Oxford Instruments) or ImageJ (NIH).

Examples of the images and videos that can be obtained with this technique are presented in Figs. 1 and 2 and Videos 1 and 2.

## 4 Notes

1. **Selection of clearing media**
   Both Ce3D and FUnGI give adequate clearing results and do not appear to destroy tissue structure or epitopes. Ce3D media is slightly more effective at clearing liver optically, is easier to work with and less viscous or adherent. However, it contains toxic substances including *N*-methylacetamide and 1-thioglycerol and as a result must be used with appropriate lab safety attire and in a fume hood. FUnGI media is less toxic and therefore may be preferable.

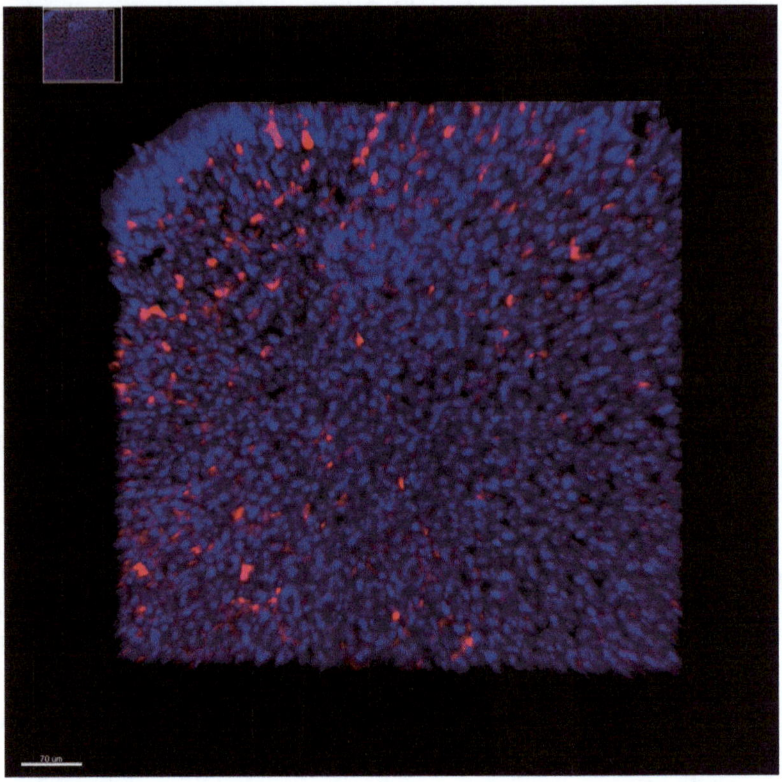

**Fig. 1** 3D rendered image of macrophages in human liver as identified by CD163 (red) and costained with DAPI to identify nuclei. Tissue cleared using Ce3D clearing media and acquired using Nikon spinning disk confocal microscope, data was rendered using the proprietary software Imaris (Oxford Instruments). Scale bar in figure represents 70 μm

2. **Sectioning tissue**

   Despite reducing optical density, sample thickness needs to be approximately 200 μm. In order to cut specimens to this thickness a vibratome sectioning device is recommended. This requires an adhesive to fix the section onto a cutting block, and a bath of PBS into which the cut sample will float and can then be retrieved before undergoing blocking and staining. Cut specimens can be stored at 4 °C in PBS until used.

3. **Notes on mounting specimens**

   The intention is to have the specimen in a refractive index matched medium, and immobile for microscopy.

   (a) Glass bottom chambers are available in many forms, with different thicknesses of glass to match microscope and optical configurations. One may have to experiment with which works optimally for each experiment, often a 1.5 glass bottom gives adequate results.

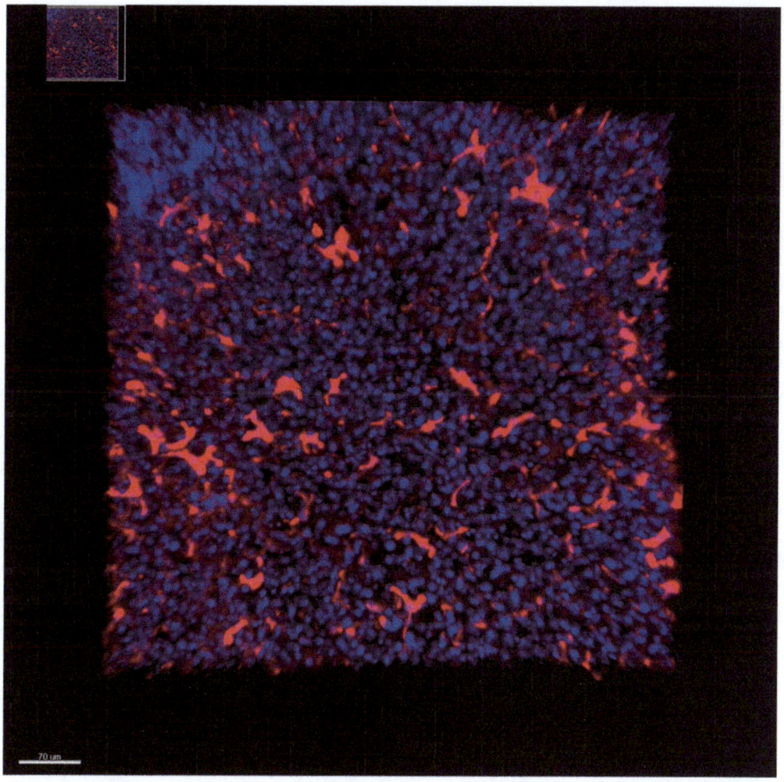

**Fig. 2** 3D rendered image of macrophages in human liver as identified by CD163 (red) and costained with DAPI to identify nuclei. Tissue cleared using FUnGI clearing media and acquired using Nikon spinning disk confocal microscope, data was rendered using the proprietary software Imaris (Oxford Instruments). Scale bar in figure represents 70 μm

(b) Try to form a closed circle when using silicone grease to form the well into which the cleared specimen will be placed. This may be best achieved by using a syringe. First, remove the syringe plunger and squeeze the grease into the syringe, reinsert the plunger and use the syringe to apply the silicone grease onto the glass bottom. Because this process can be quite awkward, it is advisable to have relative large wells in the chamber slide, such as the 6-well Zell-Kontakt imaging plate, which has chambers that are 5.1 cm in diameter.

(c) If one maintains the sample in the clearing media the refractive index should be adequately matched.

(d) If required, a further step of sealing the edges of the coverslip can be undertaken using a silicone glue such as two-component picodent twinsil (Picodent).

- Combine component A and component B and working quickly seal (it changes from a viscous liquid to a solid state quickly) the edges of the coverslip ensuring not to obscure the sample.

**Table 3**
**Composition of Ce3D clearing media**

| Volume final Ce3D buffer (mL) | *N*-methylacetamide | PBS | Histodenz (g) | Triton X-100 (μL) | 1-thioglycerol (μL) |
|---|---|---|---|---|---|
| 1 | 400 μL | 600 μL | 1.455 | 1 | 5 |
| 5 | 2 mL | 3 mL | 7.275 | 5 | 25 |
| 10 | 4 mL | 6 mL | 14.555 | 10 | 50 |

4. **Making Ce3D buffer**
The method for making the Ce3D buffer has been described by Li et al. [8]. Melted *N*-methylacetamide (Sigma) is diluted 40% (v/v) in PBS. This solution is then used to dissolve Histodenz (Sigma) to 86% (w/v) concentration [for example 1.455 g Histodenz per 1 mL 40% *N*-methylacetamide, *see* Table 3], with the mixture vortexed and incubated at 37 °C on a horizontal shaker until dissolved (this can take up to 24 h). Triton X-100 (0.1% v/v) and 1-thioglycerol (0.5% v/v) are then added to the clearing solution and further incubated at 37 °C on a horizontal shaker to dissolve. All steps must be done with appropriate lab safety equipment including gloves, a lab coat and when open to air must be performed in a chemical fume hood until mounted and sealed. *N*-methylacetamide has a melting point of 28.0 °C, therefore place container in water bath to melt. When *N*-methylacetamide melts, transfer it to chemical fume hood and work quickly as it returns to solid phase within minutes.

5. **Ce3D Buffer components**
Table 3 lists the recommended amounts and volumes to be added to make the Ce3D buffer.

6. **Making FUnGI Buffer**
The method for making FUnGI clearing media has been previously described by Rios et al. [4]. Briefly, dissolve 100 g fructose in 2.33 mL of Tris–EDTA (pH 8.0, 10×) (Sigma) and 20.97 mL $dH_2O$. Once dissolved add 110 mL glycerol. To this add 33.1 g Urea and dissolve. Once dissolved add 50 μg/mL ascorbic acid and 0.05 ng/mL L-glutathione. L-glutathione requires serial dilutions to be at a working concentration. The solution should be made in appropriately sized lab bottle (such as a 250 mL bottle) with magnetic stirrer. It is recommended to do this in a stepwise fashion, first mixing the Tris–EDTA and glycerol, then adding the fructose. Once the fructose is dissolved add the urea. It is likely to take 12–36 h to dissolve. As per Rios et al. [4], once all the components of the buffer have dissolved, they can be kept in the dark 4 °C for up to 1 month.

7. **Notes on blocking and staining buffers.**
Rios et al. [4] included Sodium dodecyl sulfate (SDS) in their wash and staining buffer. SDS does appear to improve the optical clearing of tissue, but it appears to disrupt the liver microarchitecture and antibody binding, therefore a milder detergent such as Triton X-100 is preferable.

8. **Notes on different blocking and buffers used**
For tissue blocking, Li et al. used a media with BSA and mouse serum [8], we decided to use Background Buster primarily for simplicity, but also because the protocol was described to assess mice organs, and we were investigating human liver. Rios et al. used a variety of staining and blocking agents [4]. In our experience these were not necessary, and as mentioned in **Note 6**, we found SDS to be disruptive of tissue structure.

9. **Notes on imaging**
We have used predominantly LS Confocal, and SD confocal. Lightsheet microscopy could be used in theory but we have no experience with lightsheet microscopy in samples cleared using Ce3D and FUnGI. Because of the size of specimens, it is recommended to first acquire at a lower magnification (with a 5× or 10× objective). This is orientating and can be using to focus in on areas of interest, such as portal tracts (as identified with Cytokeratin 19).

## References


1. Strauss O, Phillips A, Ruggiero K et al (2017) Immunofluorescence identifies distinct subsets of endothelial cells in the human liver. Sci Rep 7:44356
2. Wang X-N, McGovern N, Gunawan M et al (2014) A three-dimensional atlas of human dermal leukocytes, lymphatics, and blood vessels. J Investig Dermatol 134:965–974
3. Azaripour A, Lagerweij T, Scharfbillig C et al (2016) A survey of clearing techniques for 3D imaging of tissues with special reference to connective tissue. Prog Histochem Cytochem 51:9–23
4. Rios AC, Capaldo BD, Vaillant F et al (2019) Intraclonal plasticity in mammary tumors revealed through large-scale single-cell resolution 3D imaging. Cancer Cell 35(4):618–632.e6
5. Tainaka K, Murakami TC, Susaki EA et al (2018) Chemical landscape for tissue clearing based on hydrophilic reagents. Cell Rep 24:2196–2210.e9
6. Uderhardt S, Martins AJ, Tsang JS et al (2019) Resident macrophages cloak tissue microlesions to prevent neutrophil-driven inflammatory damage. Cell 177(3):541–555.e17
7. Richardson DS, Lichtman JW (2015) Clarifying tissue clearing. Cell 162:246–257
8. Li W, Germain RN, Gerner MY (2017) Multiplex, quantitative cellular analysis in large tissue volumes with clearing-enhanced 3D microscopy (C e3D). Proc Natl Acad Sci 114: E7321–E7330
9. Tainaka K, Kuno A, Kubota SI et al (2016) Chemical principles in tissue clearing and staining protocols for whole-body cell profiling. Annu Rev Cell Dev Biol 32:713–741
10. Unnersjö-Jess D, Scott L, Blom H et al (2016) Super-resolution stimulated emission depletion imaging of slit diaphragm proteins in optically cleared kidney tissue. Kidney Int 89:243–247
11. MacParland SA, Liu JC, Ma X-Z et al (2018) Single cell RNA sequencing of human liver reveals distinct intrahepatic macrophage populations. Nat Commun 9(1):4383

# Chapter 8

# Glucan-Encapsulated siRNA Particles (GeRPs) for Specific Gene Silencing in Kupffer Cells in Mouse Liver

**Michaela Tencerova**

## Abstract

Here, we describe a protocol to prepare and administer glucan-encapsulated RNAi particles (GeRPs), for specific delivery of siRNA and subsequent gene silencing in Kupffer cells (KCs) in mice. This technology is based on baker's yeast and allows gene manipulation in macrophages in a tissue-specific manner depending on the route of administration and the model that is used. GeRP administered by intravenous injection in mice are delivered to KCs. Therefore, using the GeRP technology to silence genes provides a unique method to study the function of factors expressed by KCs in the regulation of liver function.

**Key words** RNAi delivery, Intravenous administration, Kupffer cells, Mouse liver, Gene silencing, Obesity, Diabetes, Inflammation, Insulin sensitivity

## 1 Introduction

The liver is one of the largest organs in the body playing a central role in the regulation of metabolism and immune system [1, 2]. It is a heterogeneous organ, which consists of several cell types, including parenchymal cells (hepatocytes) representing the majority of liver cells (92.5%) and nonparenchymal cells, which contribute only 7% to the liver volume. In nonparenchymal fraction, sinusoidal endothelial cells (70%), Kupffer cells (KCs) (20%), hepatic stellate cells (10%), and other immune cells (e.g., dendritic cells, lymphocytes, less than 5%), which participate in the inflammatory response and regenerative processes occurring in the liver [3–5], are present. Hepatocytes play an important role in the regulation of glucose, lipid metabolism, and bile production [6, 7]. However, it has recently been shown that KCs may also contribute to the regulation of glucose metabolism independently of inflammation [8]. KCs represent the largest reservoir of resident tissue macrophages (80–90%) in the whole body, building an immune barrier in the liver. They reside in the liver sinusoids, where they come in contact with various microorganisms, toxic agents, or cell debris from

Myriam Aouadi and Valerio Azzimato (eds.), *Kupffer Cells: Methods and Protocols*, Methods in Molecular Biology, vol. 2164, https://doi.org/10.1007/978-1-0716-0704-6_8, 

circulation leading to phagocytic activity and secretion of inflammatory cytokines, chemokines, and reactive oxygen species affecting neighboring cells in the liver [9, 10].

The role of KCs in pathophysiology of liver metabolic diseases represents an attractive area of the research studying the crosstalk with other hepatic cells and their systemic effect on metabolism and inflammation [8, 11–14]. Therefore, it is important to develop techniques that can specifically manipulate KCs function but leaving the macrophages in other tissues intact. Most of the studies are using transgenic mouse models with a full-body or myeloid-specific knockout or using anti-inflammatory drugs, whose effects are not tissue or cell specific.

Here we describe a method to specifically silence genes in KCs using glucan-encapsulated RNAi particles (GeRPs), which are recognized by phagocytic cells and tissue specificity of GeRPs is based on the route of administration, that is, intravenous (i.v.) injections in mice. GeRPs-mediated RNAi silencing has been successfully shown in several studies investigating the role of tissue macrophages in the regulation of a whole body glucose metabolism and inflammation [8, 12, 15–17]. Here, we provide a detailed protocol of GeRP formulation for gene silencing, including the preparation of glucan shells (GS) from baker's yeast (*Saccharomyces cerevisiae*), the labeling of the GS with fluorescein isothiocyanate (FITC), the loading of GS with siRNA and the in vivo GeRPs administration in mice.

## 2 Materials

### 2.1 Preparation of Glucan Shells

1. SAF-Mannan yeast (*S. cerevisiae*).
2. 0.5 M sodium hydroxide solution: Add 20 g of NaOH to 1 L of deionized water. Prepare fresh.
3. Distilled water.
4. Isopropanol.
5. Acetone.
6. Erlenmeyer flask (2 L).
7. Magnetic stirring bar.
8. Hot plate with magnetic stirrer.
9. Thermometer.
10. High-speed centrifuge.
11. 250 mL centrifuge bottles.
12. 1.5 mL Eppendorf tubes.
13. Spatula.
14. Brinkman Polytron homogenizer or equivalent.

15. Fume hood.
16. 200 μM mesh filter.
17. Electric coffee grinder.
18. Hemocytometer.

### 2.2 Fluorescent Labeling of Glucan Shells

1. Glucan shells (GS).
2. Sodium carbonate buffer pH 9.2: Prepare a 0.1 M solution of sodium carbonate and a 0.1 M solution of sodium bicarbonate. To obtain 500 mL of buffer, mix 50 mL of 0.1 M sodium carbonate with 450 mL of 0.1 M sodium bicarbonate. Store at RT.
3. Fluorescein isothiocyanate (FITC) solution: 1 mg/mL FITC in absolute ethanol.
4. Absolute ethanol.
5. Filtered water.
6. Sterile water.
7. Analytical balance.
8. Erlenmeyer flask (1 L).
9. 50 mL centrifuge tubes.
10. Probe sonicator.
11. Vortex mixer.
12. Lyophilizer.

### 2.3 Loading of Glucan Shells with siRNA and Delivery to Kupffer Cells In Vivo

1. FITC-labeled GS.
2. 1 mM siRNA: dilute siRNA stock to 1 mM using 1× siRNA buffer (Dharmacon).
3. 5 mM Endo-Porter (EP; Gene Tools): Prepare a 5× solution of the 1 mM EP stock by placing the uncapped tube in a speed vacuum on high heat for approximately 3 h until completely dry. Resuspend the pellet in 200 μL RNase-free water, and heat again at 45 °C for 15 min to solubilize the dried Endo-porter (*see* **Note 1**).
4. 30 mM sodium acetate buffer: Prepare 250 mL buffer and adjust pH to 4.8 using glacial acetic acid. Sterile filter.
5. Filters (22 μm pore size).
6. PBS.
7. 1 mL syringe.
8. Vortex mixer.
9. Probe sonicator.
10. Mouse anesthesia: Isoflurane with 2% oxygen.

***2.4 Solutions for Hepatocyte and KC Immunofluorescence***

1. Hepatocyte immunofluorescence: Albumin antibody (R&D) to be used at a concentration of 25 μg/mL. This antibody has reactivity for both human and mice.
2. KC immunofluorescence: F4/80 antibody (AbD Serotec) to be used at a concentration of 1:50.
3. Collagen coated coverslips for hepatocytes: Coat coverslips using Rat Tail Collagen Type 1 (Sigma), for 6 h at room temperature or overnight at 4 °C. Aspirate the collagen solution and then dry the coverslips under UV light for 3 h (*see* **Note 2**).

## 3 Methods

Perform the preparation of glucan shells in a fume hood.

***3.1 Preparation of Glucan Shells***

1. Resuspend 100 g SAF-Mannan yeast in 1 L 0.5 M NaOH solution in a 2 L Erlenmeyer flask with a stir bar. Heat the solution to 80–85 °C for 1 h while stirring to maintain suspension. Make sure to maintain the temperature range by inserting a thermometer in the flask. Caution: hot caustic.
2. Leave the solution to cool to 40–50 °C. Then transfer the suspension into 6 × 250 mL centrifuge bottles. Centrifuge using a high-speed centrifuge at 15,000 × *g* for 10 min (*see* **Note 3**).
3. Decant off the supernatant, and resuspend the pellet in deionized water (1 L in total) using a homogenizer at 2800 × *g* (*see* **Note 4**).
4. Once the pellet has dissolved, transfer the solution back to a clean 2 L flask.
5. Add 20 g of NaOH to make a 0.5 M solution.
6. Heat to 80–85 °C for 1 h while stirring to maintain suspension. Make sure to maintain temperature range by inserting a thermometer in the flask.
7. Leave the solution to cool to 40–50 °C.
8. Transfer the suspension into the 250 mL centrifuge bottles. Centrifuge at 15,000 × *g* for 10 min. Decant off the supernatant.
9. Wash the pellets first in sterile water by adding 200 mL deionized water to each centrifuge tube. Use the homogenizer to resuspend the pellets.
10. Centrifuge at 22,000 × *g* for 15 min and decant the supernatant.

11. Repeat the wash step with water until the supernatant is clear (*see* **Note 5**).
12. Wash the pellets with isopropanol by adding 200 mL isopropanol to each centrifuge tube. Use the homogenizer to resuspend the pellets.
13. Centrifuge at 15,000 × *g* for 15 min, and decant the supernatant (*see* **Note 6**).
14. Continue to wash the pellets with isopropanol until the supernatant is clear (*see* **Note 5**).
15. Wash the pellets in acetone by adding 200 mL acetone to each of the six tubes. Use the homogenizer to resuspend the pellets.
16. Centrifuge at 15,000 × *g* for 20 min and decant the supernatant. Repeat the wash with acetone twice.
17. Centrifuge the suspension at 15,000 × *g* for 30 min to collect the glucan shells. Decant the supernatant.
18. Break and mix the packed pellet using a spatula, and allow the acetone to evaporate off under a fume hood overnight.
19. Grind the resulting pellet in a coffee grinder into a fine powder.
20. Filter the powder through a 200 μm mesh filter, and transfer into tubes for storage at −20 °C.
21. The yield can be determined with the use of a hemocytometer (*see* **Note** 7).
22. Measurement of hydrodynamic volume: to measure the volume that can be absorbed by the GS, weigh 10 mg of glucan shells in a 1.5 mL Eppendorf tube. Record the weight of the tube containing the GS, and then resuspend the GS in 400 μL water by vortexing. Allow to incubate at RT for 1 h; vortex every 10 min. Thereafter, pellet the GS by centrifugation at 9000 × *g* for 10 min, remove excess water, and weigh the tube again. The hydrodynamic volume is the difference in weight before and after the addition of water divided by the weight of the dry GS (mg water/mg GP) and should be around 20–30 μL per mg of GS.

### 3.2 Fluorescent Labeling of Glucan Shells

Glucan shells can be labeled with different fluorophores; here we describe the process of how to label them with FITC (*see* **Note 8**).

1. Wash glucan shells with sodium carbonate buffer by adding 100 mL of sodium carbonate buffer to 1 g of GS in an Erlenmeyer flask.
2. Centrifuge at 3000 × *g* for 10 min.
3. Remove the supernatant, and resuspend the particles in fresh 100 mL of sodium carbonate buffer.

4. Add 10 mL of FITC solution to the buffered glucan shell suspension.
5. Stir at room temperature for ~16 h protected from light.
6. Transfer the solution to 50 mL centrifuge tubes.
7. Centrifuge at 4700 × *g* for 5 min.
8. Wash with 40 mL filtered water per tube. Resuspend the particles by thoroughly vortexing the tubes, and then sonicate until solution is homogenous.
9. Centrifuge at 4700 × *g* for 5 min and remove the supernatant.
10. Repeat the wash until the supernatant is clear (*see* **Note 9**).
11. Wash the particles with 40 mL absolute ethanol.
12. Rehydrate the particles by resuspending the pellet in 40 mL sterile water inside a tissue culture hood.
13. Centrifuge at 4700 × *g* for 10 min and remove the supernatant.
14. Add 10 mL sterile water in the hood.
15. Vortex thoroughly and sonicate until solution is homogenous.
16. Freeze the particles in liquid nitrogen and dry using a lyophilizer.
17. Store at −20 °C protected from light.

### 3.3 Loading of Glucan Shells with siRNA and Delivery to Kupffer Cells by Intravenous Injections In Vivo

1. Incubate acetate buffer, siRNA, and Endo-porter at RT for 15 min according to the volumes described in Table 1 (*see* **Notes 10–13**).
2. Add the siRNA–EP solution from **step 1** to the FITC-labeled glucan shells according to the ratio in Table 1.
3. Carefully vortex to mix and incubate at RT for 1 h in the dark.
4. Add PBS (see volume in Table 1), and sonicate for 5 × 10 s at 30% amplitude.

**Table 1**
**Reagents for loading glucan shells with siRNA**

| | For 1 dose | Volume (μL) |
|---|---|---|
| 1 mM siRNA | 3 cnmol | 3 |
| 5 × EP | 50 nmol | 10 |
| Glucan shells | 1 mg | – |
| 30 nM acetate buffer, pH 4.8 | – | 13 |
| PBS | – | 974 |

The following amounts are for the preparation of GeRPs for 5 × 200 μL injections

5. Administer the glucan-encapsulated RNAi particles by intravenous (i.v.) injections every second day for 5–10 days with a volume of 200 μL. Vortex the suspension before each injection (*see* **Notes 14 and 15**). For i.v. injection the mice are anesthetized with isoflurane–2% oxygen. I.v. Injections of obese mice with fluorescent labeled particles will result in an accumulation of GeRPs in the liver, specifically in KCs but not hepatocytes or other liver cells (Figs. 1 and 2).
6. Mice are sacrificed and tissues harvested 24–48 h after the last injection.

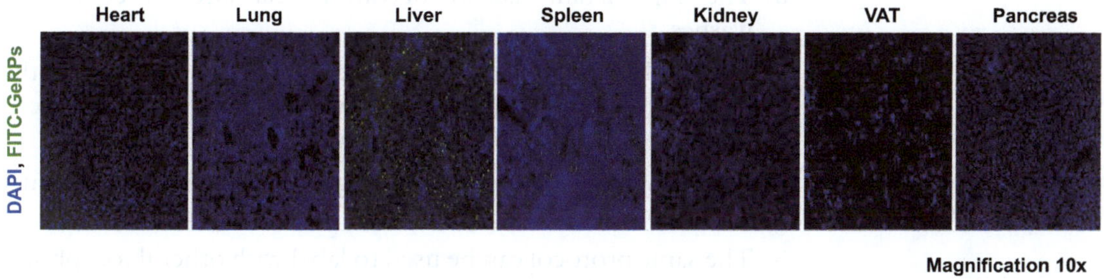


**Fig. 1** Biodistribution of intravenously injected FITC-labeled GeRPs in mice. FITC-labeled GeRPs are localized in liver, in Kupffer cells (KCs) without targeting macrophages in other tissues including the heart, lung, spleen, kidney, adipose tissue, or pancreas. Twenty-four hours after administration of GeRPs, tissues were collected, fixed, paraffin-embedded, and sectioned. The fluorescence of the FITC-GeRPs was evaluated on unstained sections. All images were acquired with a Zeiss Observer Z1 microscope (×10 magnification)

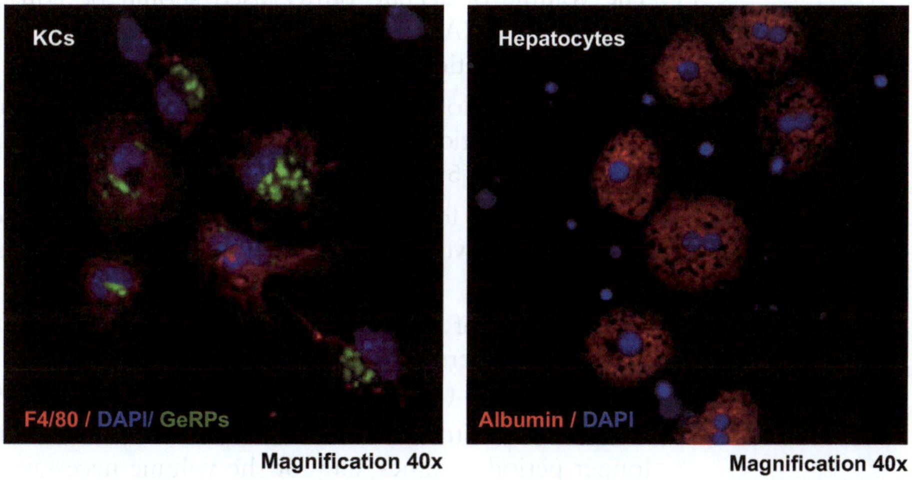


**Fig. 2** Cell-specific delivery of FITC-labeled GeRPs. Localization of FITC-labeled GeRPs in Kupffer cell (KCs) but not in hepatocyte fractions. Cells were isolated after 24 h i.v. administration of FITC-labeled GeRPs from the liver of 14-week-old male C57BL/6J mouse on a regular chow diet. KCs (left panel) were stained with an antibody against F4/80 (red) and hepatocytes (right panel) with an antibody against albumin (red). Nuclei were stained with DAPI (blue) (×40 magnification)

## 4 Notes

1. EP is an amphipathic peptide composed of leucine and histidine residues used to bind the siRNA to be loaded into glucan shells.
2. Coated coverslips can be coated ahead of time and stored at 4 °C sealed in Parafilm M®.
3. We use a Beckman Avanti JXN-30 with a 250 mL flask rotor.
4. When using the homogenizer, the solution can foam, so make sure to start with 1 L of water for the washing steps so that you know how much water is added to the pellets.
5. The supernatant usually becomes clear after three to five washes.
6. Optional stopping point: The protocol can be stopped after the first wash with isopropanol, and bottles containing the glucan shell pellet can be left at 4 °C overnight.
7. With a starting amount of 100 g of yeast, the final product should be around 6–8 g GS with $1 \times 10^9$ GS per mg.
8. The same protocol can be used to label with other fluorophores (e.g., rhodamine).
9. The supernatant usually becomes clear after five to eight washes.
10. Depending on the efficiency of the sequence, the amount of siRNA used per mg of GS can be varied from 1 to 5 nM. The efficiency is determined by performing a dose response.
11. The volume of acetate buffer used should be equal to the volume of siRNA and EP combined and needs to be adjusted if the concentration of siRNA is altered.
12. It is possible to load the GS with multiple siRNAs as long as the final concentration does not exceed the maximum amount recommended (5 nM per mg of GS).
13. Prepare GeRPs for one extra mouse for every five mice since small amounts will be lost each time when loading the syringes for injection.
14. The duration of the treatment depends on the target. Gene silencing is generally observed between 5 and 10 every second day intravenous (retro orbital) injections with 200 μL GeRPs.
15. GeRPs can be stored at 4 °C for up to 1 week if needed. For longer period of times, aliquot the volume necessary for one daily dose, flash-freeze, and store at −20 °C. GeRPs cannot be stored at −20 °C without flash freezing first, otherwise they form aggregates that are difficult to break down.

## Acknowledgments

This work was supported by research grants from the Czech Science Foundation GACR 20-03586S and the START UP program from Institute of Physiology, the Czech Academy of Sciences.

### References


1. Bogdanos DP, Gao B, Gershwin ME (2013) Liver immunology. Compr Physiol 3 (2):567–598. https://doi.org/10.1002/cphy.c120011
2. Robinson MW, Harmon C, O'Farrelly C (2016) Liver immunology and its role in inflammation and homeostasis. Cell Mol Immunol 13(3):267–276. https://doi.org/10.1038/cmi.2016.3. Epub 2016 Apr 11
3. Smedsrod B, Pertoft H, Gustafson S, Laurent TC (1990) Scavenger functions of the liver endothelial cell. Biochem J 266(2):313–327
4. Knook DL, Sleyster EC (1976) Separation of Kupffer and endothelial cells of the rat liver by centrifugal elutriation. Exp Cell Res 99 (2):444–449
5. Kmieć Z (2001) Cooperation of liver cells in health and disease. Adv Anat Embryol Cell Biol 161:III–XIII, 1–151
6. Wang X, Sato R, Brown MS, Hua X, Goldstein JL (1994) SREBP-1, a membrane-bound transcription factor released by sterol-regulated proteolysis. Cell 77(1):53–62
7. Jungermann K, Thurman RG (1992) Hepatocyte heterogeneity in the metabolism of carbohydrates. Enzyme 46(1–3):33–58
8. Morgantini C, Jager J, Li X, Levi L, Azzimato V, Sulen A, Barreby E, Xu C, Tencerova M , Näslund E, Kumar Ch, Verdeguer F, Straniero S, Hultenby K, Björkström NK, Ellis E, Rydén M, Kutter C, Hurrey T, Lauschke V, Boucher J, Tomčala A, Krejčová G, Bajgar A, Aouadi M (2019) Liver macrophages regulate metabolism independently of inflammation. Nat Metab. 1, 497. Published 25 Mar 2019
9. Jager J, Aparicio-Vergara M, Aouadi M (2016) Liver innate immune cells and insulin resistance: the multiple facets of Kupffer cells. J Intern Med 280(2):209–220. https://doi.org/10.1111/joim.12483. Epub 2016 Feb 11
10. Baffy G (2009) Kupffer cells in non-alcoholic fatty liver disease: the emerging view. J Hepatol 51(1):212–223. https://doi.org/10.1016/j.jhep.2009.03.008
11. Stienstra R, Saudale F, Duval C, Keshtkar S, Groener JE, van Rooijen N, Staels B, Kersten S, Muller M (2010) Kupffer cells promote hepatic steatosis via interleukin-1beta-dependent suppression of peroxisome proliferator-activated receptor alpha activity. Hepatology 51 (2):511–522. https://doi.org/10.1002/hep.23337
12. Tencerova M, Aouadi M, Vangala P, Nicoloro SM, Yawe JC, Cohen JL, Shen Y, Garcia-Menendez L, Pedersen DJ, Gallagher-Dorval-K, Perugini RA, Gupta OT, Czech MP (2015) Activated Kupffer cells inhibit insulin sensitivity in obese mice. FASEB J 29(7):2959–2969. https://doi.org/10.1096/fj.15-270496
13. Tosello-Trampont AC, Landes SG, Nguyen V, Novobrantseva TI, Hahn YS (2012) Kuppfer cells trigger nonalcoholic steatohepatitis development in diet-induced mouse model through tumor necrosis factor-alpha production. J Biol Chem 287(48):40161–40172. https://doi.org/10.1074/jbc.M112.417014
14. Petrasek J, Bala S, Csak T, Lippai D, Kodys K, Menashy V, Barrieau M, Min SY, Kurt-Jones EA, Szabo G (2012) IL-1 receptor antagonist ameliorates inflammasome-dependent alcoholic steatohepatitis in mice. J Clin Invest 122 (10):3476–3489. https://doi.org/10.1172/jci60777
15. Aouadi M, Tesz GJ, Nicoloro SM, Wang M, Chouinard M, Soto E, Ostroff GR, Czech MP (2009) Orally delivered siRNA targeting macrophage map 4k4 suppresses systemic inflammation. Nature 458(7242):1180–1184. https://doi.org/10.1038/nature07774
16. Aouadi M, Tencerova M, Vangala P, Yawe JC, Nicoloro SM, Amano SU, Cohen JL, Czech MP (2013) Gene silencing in adipose tissue macrophages regulates whole-body metabolism in obese mice. Proc Natl Acad Sci U S A 110(20):8278–8283. https://doi.org/10.1073/pnas.1300492110
17. Aouadi M, Vangala P, Yawe JC, Tencerova M, Nicoloro SM, Cohen JL, Shen Y, Czech MP (2014) Lipid storage by adipose tissue macrophages regulates systemic glucose tolerance. Am J Physiol Endocrinol Metab 307(4): E374–E383. https://doi.org/10.1152/ajpendo.00187.2014

# Chapter 9

# Engraftment of Human Hepatocytes in the PiZ-NSG Mouse Model

**Qiushi Tang, Gwladys Gernoux, Yangfun Cheng, Terence Flotte, and Christian Mueller**

## Abstract

We recently described the generation of a novel mouse strain that efficiently and readily engrafts human primary hepatocytes to create liver xenografts (Borel et al., Mol Ther, 25: 2477–89, 2017). A transgenic mouse strain expressing a human PiZ allele for the SerpinA1 gene was crossed with the NOD-SCID-gamma chain knockout (NSG) strain to create a recipient strain (PiZ-NSG) for human hepatocyte xenotransplantation. In this chapter we provide a description of the methods to achieve these liver xenografts in the PiZ-NSG mouse.

**Key words** Hepatectomy, Human hepatocyte, Engraftment, Intrasplenic injection, Humanized mouse model, Liver xenograft, Liver regeneration, Hepatocyte isolation

## 1 Introduction

Alpha-1 antitrypsin (AAT) deficiency is mainly caused by common missense mutation (E342K, known as the PiZ allele) which results in impaired secretion of AAT from the liver. In the liver this mutation causes the accumulation of Z-mutant AAT protein and triggers hepatocyte injury leading to inflammation and cirrhosis. We have previously shown that the cell death and regeneration cycle imparted by the PiZZ burdened livers will allow normal donor human hepatocytes or genetically corrected murine hepatocytes to progressively outcompete host hepatocytes [1]. Therefore, the (NSG)-PiZ transgenic mice expressing mutant Z-AAT on the severe immune-deficient NSG background allows for the repopulation of the liver by normal donor human hepatocytes. In this chapter we describe the surgical methods (Fig. 1) necessary for engrafting these mice, as well as the ELISA assays to monitor engraftment and finally we describe an FACS-based protocol for the separation of human and mouse hepatocytes.

Myriam Aouadi and Valerio Azzimato (eds.), *Kupffer Cells: Methods and Protocols*, Methods in Molecular Biology, vol. 2164, https://doi.org/10.1007/978-1-0716-0704-6_9, 

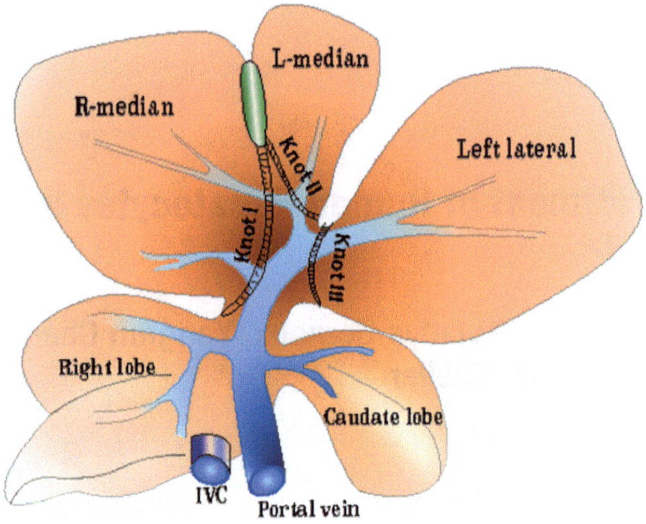


**Fig. 1** Liver structure. Three knots (knot I, II, and III) are made to tie R-median, L-median, and left lateral lobes before excision

## 2 Materials

1. The materials and tools are listed below. Common animal surgery tools and anesthesia equipment and supplies such as scissors, forceps, and needle holders are not listed here.
2. Ring Tipped Forceps, 10 cm, Straight, 4.8 mm IDWPI.
3. Ethicon Suture, 0, Silk black braided A306.
4. 5-0 Vicryl suture absorbable (Ethicon).
5. Vetbond™ Tissue Adhesive (Santa Cruz animal health).
6. Human hepatocytes (Bioreclamation IVT).
7. Hank's Balanced Salt Solution (HBSS) (Sigma).
8. Hair removal cream (Nair CVS).
9. Buprenorphine SR-LAB (ZooPharm).
10. Neo Poly Dex Ophthalmic Ointment (Allivet Pet Pharmacy).
11. Germinator 500 Glass Bead Sterilizer (CellPoint Scientific).
12. Human Albumin ELISA Quantitation Set (E80-129, Bethyl laboratories).
13. High binding 96-well plates (Immulon4, Dynatech Laboratories).
14. HRP-conjugated antibody, 3,3′,5,5′-tetramethylbenzidine (TMB) peroxidase substrate (KPL).
15. ELISA stop solution 2 N $H_2SO_4$ (Fisher Scientific).
16. Citra (BioGenex).

17. Background Sniper (Biocare Medical).
18. Rabbit anti-human albumin (ab2604, Abcam).
19. Mach2 Goat × Rabbit HRP polymer (RHRP520, Biocare Medical).
20. 3,3′-diaminobenzidine (DAB) chromogen (Biocare Medical).
21. Churukian-Allison-Tacha (CAT) Hematoxylin (Biocare Medical).
22. VersaMax microplate reader (Molecular Devices).
23. Liver perfusion medium (17701038, Life Technologies).
24. Liver digest medium (17703034, Life Technologies).
25. Hepatocyte wash medium (17704024, Life Technologies).
26. Falcon® 70 μm Cell Strainer (Corning).
27. Percoll (GE Healthcare).
28. 5 ml polystyrene round-bottom tubes (352052, Falcon)
29. Phosphate-Buffered Saline (PBS) (21-040-CM, Corning).
30. Anti-mouse/human CD324 (DECMA-1, Biolegend).
31. Anti-HLA-ABC (G46-2.6, BD Biosciences).
32. FACS-LSRII flow cytometer (BD Biosciences).
33. FlowJo software (Tree Star Inc.).
34. Minimum Essential Media (MEM) (12360-038, ThermoFisher).
35. Human Serum—Type AB (S40110H, Atlanta Biologicals).

## 3 Methods

### *3.1 Presurgical Instrument and Equipment Preparation*

1. Use sterile single-use (disposable) or sterilize all materials and substances used for surgery or placed at the side of the animals. In case of multiple surgeries on the same day, use a bead sterilizer between animal surgeries.
2. Perform surgeries on a thermoregulated operating board designed to maintain a temperature of 37 °C throughout the procedure, which routinely takes 15–20 min to perform.
3. Keep mice on a heating blanket or under a heating lamp during the surgery to maintain body temperature.

### *3.2 Surgeon's Presurgical Preparation*

1. Wear clean lab coat, face mask, and scrubs.
2. Perform a hand wash with disinfectant scrub before putting on sterile gloves.
3. Place a sterile drape on operative site to prevent any contamination of the incision sites.

### 3.3 Animal Presurgical and Anesthetic Preparation (See Note 1)

1. Remove hair from the surgical area (Fig. 2a). The area must be at least twice the size of the surgery area you will need. Surgery preparation should take place in a separate location than where the surgical operation will be performed or 1 day before the surgery.
2. Place the recipient animals in closed acrylic ventilation chamber hooked up to an anesthesia machine that mixes the 2–4% isoflurane with 1 l/min of oxygen. Carbon traps are attached to the acrylic box and collect any access isoflurane (*see* **Note 2**).
3. Transfer the unconscious animal to the surgery site.
4. Maintain anesthesia using a nose cone continued inhalation.
5. Perform a toe-pinch reflex test every 5 min to test that the mouse is truly unconscious under isoflurane.
6. Monitor respiration under anesthesia to assess whether the mouse is deep enough for surgery. Isoflurane dose will be increased during surgical procedure if the mouse is positive to gentle toe pinch.
7. Apply ophthalmic ointment to both eyes of the mouse to prevent desiccation or any further irritation while under isoflurane–oxygen.
8. Clean the surgery area with alcohol. Repeat three times.
9. Disinfect mouse skin with Betadine in a limited area (Fig. 2b). Application to a larger area can cause potential hypothermia.

### 3.4 Partial Liver Hepatectomy

1. Make a small median-line incision just large enough to expose the surgical area (*see* **Note 3**) (Fig. 2c, d).
2. Push out three lobes of the liver: the right and left median and left lateral from the animal cavity (Fig. 3).
3. Tie the median lobe of the liver and the left lateral lobe using PERMA-Hand silk #0 black braided strands (Fig. 4a–c). This area represents 65–70% of the total liver.
4. Excise the lobes (Fig. 4d).
5. Place the remaining liver back into the abdominal cavity.
6. Close the abdomen in two layers. Use 5-0 Vicryl suture in simple continuous method for the abdominal muscle layer and Vetbond™ or a similar surgical adhesive for the skin layer.

### 3.5 Human Hepatocytes Transplantation Procedure

1. During the same surgery, human hepatocytes are transplanted into the remaining liver immediately after the partial hepatectomy by intrasplenic injection. In this protocol, frozen human hepatocytes are used and procedure is detailed in **Note 4**.
2. Make a second incision in the flank and expose the spleen (Fig. 5a).

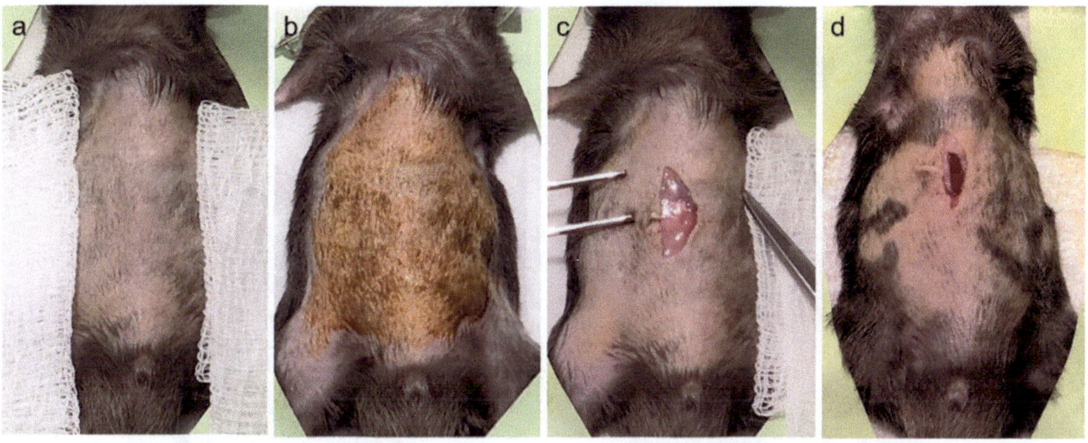


**Fig. 2** Animal skin preparation for abdominal incision. An area representing at least twice the surface of the surgical area is shaved (**a**). The surgical area is disinfected using alcohol and betadine (**b**) and a median-line incision (less than 1 cm) is made (**c**). The second layer cut is smaller than the skin layer (**d**)

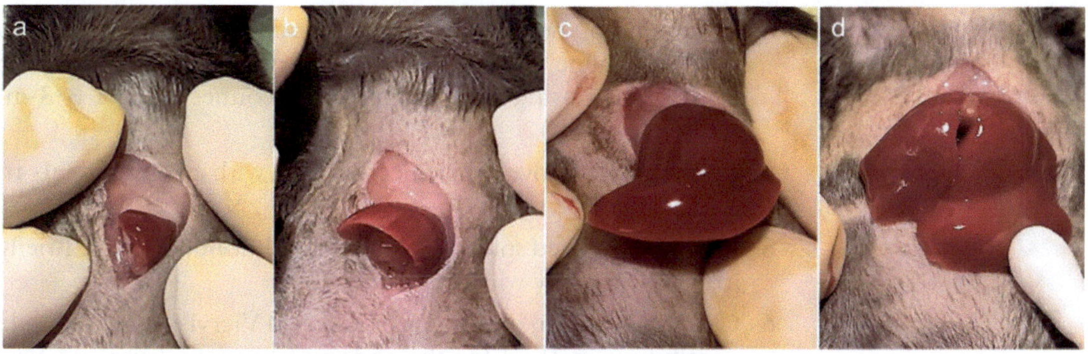


**Fig. 3** Liver exposition after abdominal incision. After incision, the three liver lobes (R-median, L-median, and left lateral) are gently pushed out of the body (**a–d**). The small opening on skin will hold the liver tied from going back into the cavity

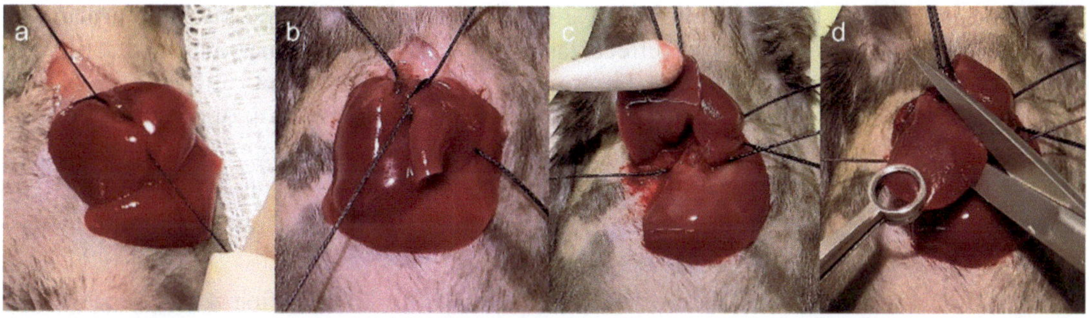


**Fig. 4** Liver lobes ligature and excision. The R-median lobe (**a**) and L-median lobe (**b**), and the Left lateral lobe (**c**) are tied using PERMA-Hand silk #0 black braided before performing lobe excision (**d**)

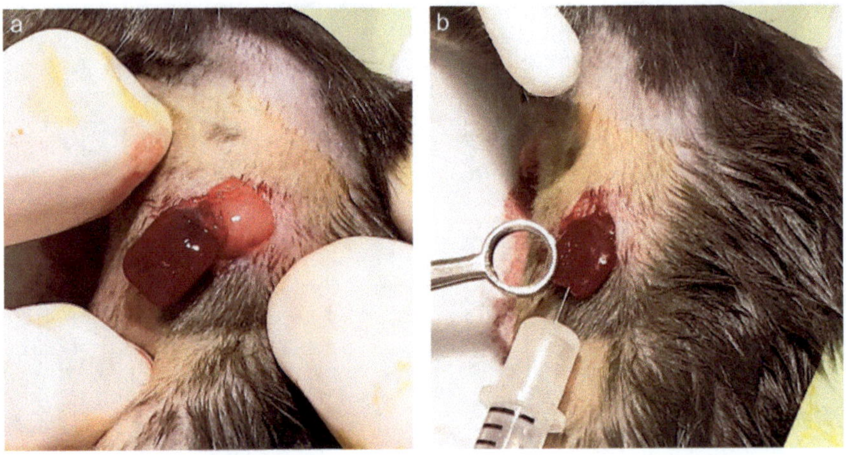


**Fig. 5** Spleen exposition and hepatocyte injection. For surgery, the animal is placed in a lateral recumbency position. After disinfecting the surgical area, an incision is made to expose the spleen (**a**). One million hepatocytes in 50 μl of Hank's balanced solution are then injected into the spleen (**b**)

3. Resuspend one million human hepatocytes in 50 μL of Hank's Balanced Saline solution and inject them into the inferior pole of the spleen using a 25-gauge needle connected to a 1/3-ml syringe (Fig. 5b).
4. Apply pressure over the injection site with a sterile swab to achieve hemostasis.
5. Place the exposed spleen back into the abdominal cavity. Briefly irrigate the cavity with sterile saline.
6. Close the incision in two layers. Use 5-0 Vicryl suture in simple continuous method for the abdominal muscle layer and Vetbond™ or a similar surgical adhesive for the skin layer. Surgical adhesive leads to a less chewing on the wound from the animals and a better healing.

### 3.6 Animal After Surgery Care

1. Administer 1.2 mg/kg of Buprenorphine SR-LAB subcutaneously (effects last for 3 days) to all animals before starting the surgery. The surgery is usually short and can be accomplished within 5–10 min, so the thermoregulation of the animal during surgery will not be critical.
2. Monitor all the animals postoperatively until they maintain sternal recumbence, which after being under isoflurane anesthesia typically occurs within 2–5 min.
3. Monitor the mice every hour for the first 4 h to ensure that they are eating and drinking as usual. Postsurgical symptoms may include guarding of the surgical site, vocalization, reclusive behavior, and loss of appetite. If any animal is experiencing any abnormal discomfort, pain or losses of more than 15% of its body weight, euthanasia will be considered in consultation with a veterinarian.

### 3.7 Liver Perfusion/Mouse Primary Hepatocyte Isolation

1. Euthanize the mice using a dioxide ($CO_2$) overdose and cervical dislocation.
2. Make an abdominal incision to expose the abdominal cavity. The viscera are arranged to expose the inferior vena cava.
3. Use a catheter needle to puncture the inferior vena cava and insert ~1 cm into the inferior vena cava.
4. Apply suture silks to secure the catheter.
5. Attach pump tubing at the end of the catheter. The portal vein is sliced to allow outflow. The diaphragm is open.
6. Clamp the inferior vena cava above the liver.
7. Perfuse the liver with 50 ml of liver perfusion medium pumped through the circulatory system, followed by 50 ml of liver digest medium.
8. Remove the liver and immerse it in ice-cold hepatocyte wash medium (HWM) allowing for the hepatocytes to diffuse into solution.
9. Collect the cell suspension and pass through a 70 μm mesh.
10. Centrifuge at 50 g for 3 min at 4 °C.
11. Add 10 ml of HWM and 10 ml 90% Percoll in phosphate-buffered saline (PBS; $Ca^{2+}$, $Mg^{2+}$ free) to the cells to remove nonviable cells.
12. Wash the hepatocyte cell suspension with 30 ml of HWM. Repeat two times.

### 3.8 Evaluation of the Engraftment

#### 3.8.1 Human Albumin ELISA

1. Collect mouse serum biweekly pre- and postengraftment.
2. Determine human albumin levels postengraftment using the human albumin ELISA quantitation kit (Fig. 6).
3. Samples and standard are diluted with blocking solution (PBS–3% BSA).
4. Add 100 μl of diluted coating antibody (1:100) to each well using high binding 96-well plates. Run each standard or sample in duplicate.
5. Incubate at room temperature (20–25 °C) for 1 h.
6. Wash plate five times with Wash buffer (PBS–0.5% Tween).
7. Add 200 μl of Blocking Solution to each well. Incubate at room temperature for 30 min.
8. Wash plate five times with Wash buffer.
9. Add 100 μl of standard or diluted samples to well. Incubate at room temperature for 1 h.
10. Wash plate five times with Wash buffer.
11. Add 100 μl of diluted HRP detection antibody (1:5000) to each well. Incubate at room temperature for 1 h.

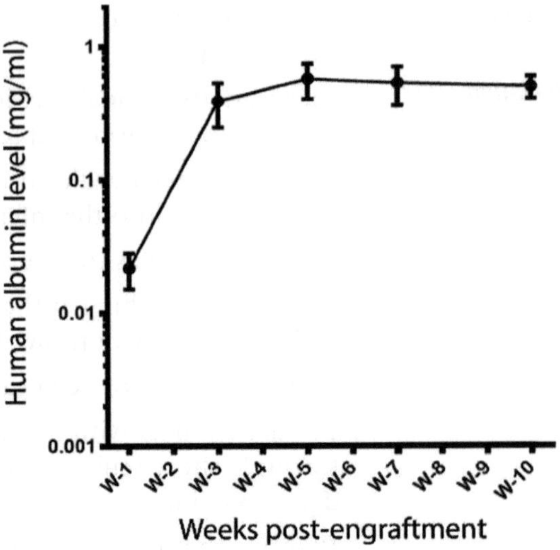


**Fig. 6** Human albumin levels in NSG-PiZ mice after hepatocyte engraftment. Serum was collected biweekly for 10 weeks after one million human hepatocyte engraftment. Human albumin levels were measured by ELISA ($N = 6$)

12. Wash plate five times.
13. Add 100 μl of TMB Substrate Solution to each well.
14. Develop the plate in the dark at room temperature for 15 min.
15. Stop reaction by adding 100 μl of Stop Solution to each well.
16. Measure absorbance on a plate reader at 450 nm (VersaMax microplate reader).

*3.8.2 Human Albumin Staining Immunohistochemistry*

1. Embed liver block in paraffin (all steps are carried out at room temperature, unless indicated otherwise).
2. Make 4 μm serial sections of mouse livers.
3. Deparaffinize and rehydrate sections (*see* **Note 5**).
4. Treat with Citra at 98 °C for 30 min.
5. Apply background Sniper to reduce nonspecific background staining.
6. Incubate sections with rabbit anti-human albumin (1:4000) for 2 h.
7. Wash five times by incubating for 5 min in PBST.
8. Stain with Mach2 Goat × Rabbit HRP polymer, the 3,3′-diaminobenzidine (DAB) chromogen and Churukian-Allison-Tacha (CAT) hematoxylin counterstain.
9. Wash five times by incubating for 5 min in PBST.
10. Use Cytoseal to coverslip the slide.

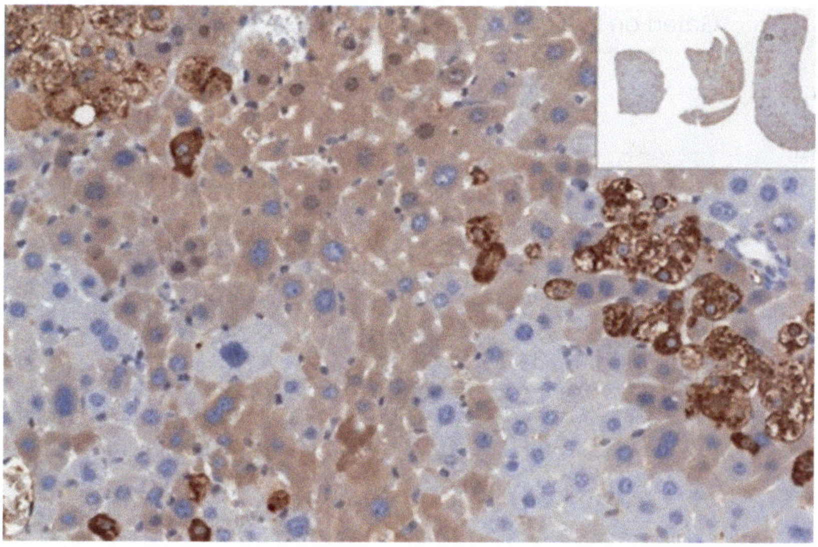

**Fig. 7** Human albumin staining in NSG-PiZ mice engrafted with human hepatocytes. Liver tissue were harvested at 14 weeks and stained for human albumin

11. Image is scanned by Aperio scanner from a Leica microscope (Fig. 7).

*3.8.3 Flow Cytometry Assay*

1. Stain one million hepatocytes on ice for 30 min using anti-mouse/human CD324 (1:20) and anti-HLA-ABC antibodies (1:5) in 50 μl PBS 1×.
2. Add 100 μl of 1× PBS and centrifuge at 300 × *g* for 5 min at 4 °C. Repeat twice.
3. Resuspend cell pellet in 200 μl of 1× PBS supplemented with 0.5% FBS and 2 mM EDTA.
4. Acquire cells using FACS-LSRII flow cytometer.
5. Analyze cell population using FlowJo software (Tree Star, Inc). Hepatocytes are gated as CD324+ cells, and human hepatocytes are defined as cells expressing the human HLA-ABC marker within the CD324+ subset (Fig. 8).

## 4 Notes

1. All surgeries follow the guidelines indicated for rodent surgery on the IACUC Animal Care and Use handbook. Aseptic techniques and sterile instruments are used during all procedures. US government policy requires the use of pharmaceutical grade drugs especially for anesthetics or analgesics. Exceptions include unavailability of pharmaceutical grade drug and lack of a suitable alternative. Exceptions require scientific

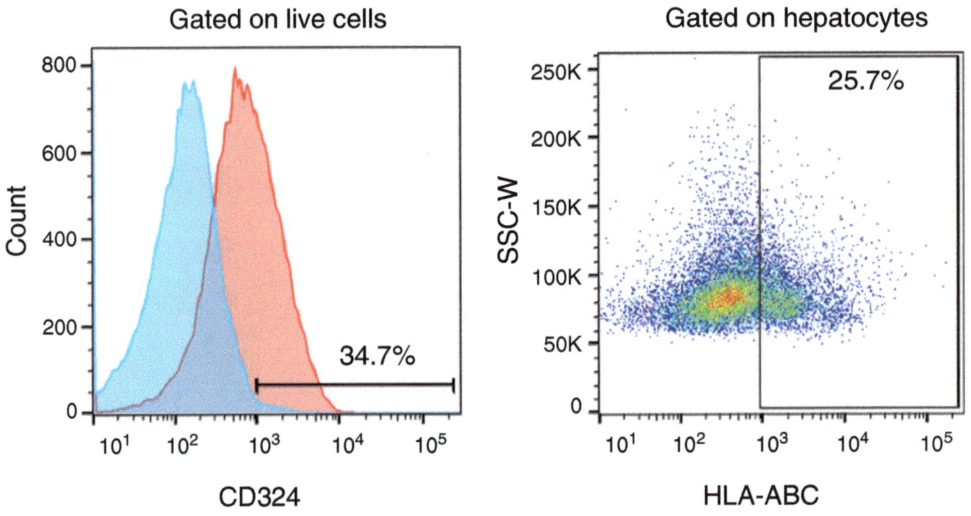


**Fig. 8** Hepatocytes of human origin quantified by flow cytometry. Hepatocytes were gated as CD324+ cells and human hepatocytes are defined as cells expressing the human HLA-ABC marker within the CD324+ subset

justification. Respiratory gating is investigated and utilized for all procedures requiring anesthesia.

2. Carbon traps are weighed before and after each use and are disposed in biohazard bins and replaced after weight increases by 50 g.
3. For partial hepatectomy procedure, a median-line incision is made as small as possible: less than 1 cm.
4. Processing Frozen Hepatocytes: Prewarm MEM medium plus 1% human serum type AB at 37 °C and transfer 50 ml of warm medium to a sterile 50 ml tube. Immediately immerse the vial of cells into a 37 °C water bath. Once thawed, add the cells to 30 ml of warm MEM medium supplemented with 1% of human serum type AB and gently invert the tube three times. Centrifuge the cells at 50 g for 5 min at room temperature. Discard the supernatant. Resuspend the pellet by gently tapping the base of the centrifuge tube. Add 5 ml of HBSS. Determine number of viable cells using trypan blue. Centrifuge the cells at 50 g for 5 min at room temperature. Dilute to the desired concentration using HBSS.
5. Deparaffinize and rehydration sections. Treat the section with following steps: 10 min incubation in xylene, repeat twice; 2 min incubation in 100% alcohol, repeat twice; 2 min incubation in 95% alcohol, repeat twice; 10 min incubation in 3% $H_2O_2$; wash three times in water.

## Acknowledgments

The mouse liver diagram in Fig. 1 was created by Mr. Hong Ma, Gene Therapy Center University of Massachusetts Medical School.

## Reference

1. Borel F, Tang Q, Gernoux G, Greer C, Wang Z, Barzel A et al (2017) Survival advantage of both human hepatocyte xenografts and genome-edited hepatocytes for treatment of alpha-1 antitrypsin deficiency. Mol Ther 25 (11):2477–2489

Chapter 10

# Kupffer Cell Characterization by Mass Cytometry

**Camille Blériot, Shamin Li, Muhammad Faris Bin Mohd Kairi, Evan Newell, and Florent Ginhoux**

## Abstract

Kupffer cells are the liver-resident macrophages lining the sinusoids and are mostly known for their role of scavengers, as crucial keepers of organ integrity. But due to the many fundamental functions of the liver notably linked to detoxication, metabolism, protein synthesis, or immunology, Kupffer cells are exposed to a dynamic environment and constantly adapt themselves by modulating their gene and protein expressions. In this context, the characterization of these cells at steady-state and upon challenges may be limited by the classical microscopy or flow cytometry which allow for the use of only few selected markers. On the other end, transcriptomic approach, although being very powerful, can be costly and time-consuming. So mass cytometry offers a good compromise, allowing for the monitoring of a representative set of markers (up to 40) in a simple experiment. Herein, we describe a straightforward experimental and analysis workflow for Kupffer cell characterization by mass cytometry.

**Key words** Immunology, Mass cytometry, CyTOF, Macrophages, Kupffer cells

## 1 Introduction

Flow cytometry is still the gold standard technology widely used in almost all immunology laboratories. It is based on antibodies coupled to fluorophores and is therefore limited by the spectral overlap between the signals from the different fluorophores used in the panel. This overlap imposes the manual and so potentially biased compensation of the signals and limits the number of markers measured simultaneously. For this reason, the vast majority of analyses do not exceed ten markers. More recently, RNA sequencing technologies notably at the single cell level have been developed and are revolutionizing our understanding of cell biology. This approach provides a complete overview of what a cell is doing at a defined timepoint, the major limitation being the cost and the time needed between sample processing and data analysis.

Mass cytometry, often abbreviated as CyTOF, though this name initially was used for the first machine designed and

Myriam Aouadi and Valerio Azzimato (eds.), *Kupffer Cells: Methods and Protocols*, Methods in Molecular Biology, vol. 2164, https://doi.org/10.1007/978-1-0716-0704-6_10, 

commercialized by Fluidigm, was described in a seminal paper 10 years ago [1]. The aim was to keep the simplicity of utilization of the flow cytometry but to circumvent the problem of spectral overlap by using metals instead of fluorophores. By coupling the antibodies with stable isotopes with incremental atomic number z, it was possible to design a panel of 20 markers that does not require any manual compensation. Nowadays, machines permit the use of 40 markers simultaneously allowing for the study of the immune landscape in lung adenocarcinoma [2] or the heterogeneity of human innate lymphoid cells [3] for example.

Resident tissue macrophages (RTMs) are immune scavengers maintaining tissue homeostasis by engulfing apoptotic cells and potential invaders. Most often, they colonize the tissues during embryonic development and exhibit therefore a very high level of functional adaptations to their niche of residence [4]. One among many examples is the Kupffer cells which populate liver sinusoids and represent one of the most abundant population of RTMs [5]. The liver is a vital organ, responsible of many different functions and is therefore affected by a lot of diseases [6]. The precise role of Kupffer cells in the establishment and development or control of liver diseases is far from being totally understood. Furthermore, the liver is not so homogeneous, and labor is divided into several functional zones [7]. This intrinsic heterogeneity, coupled to the constant changes in the local microenvironment in healthy or diseased livers, affect Kupffer cell biology. In this context, mass cytometry and its advantages stand out as a very promising alternative between limited flow cytometry and time-consuming transcriptomics. Herein, we will provide a workflow starting from Kupffer cell isolation to the final analysis of mass cytometry data (Fig. 1). Of note, even if this workflow has been efficiently used for Kupffer cell characterization, it should be possible to use it to study other immune cell populations by adapting the analysis panel.

## 2 Materials

### 2.1 Reagents

1. RPMI complete medium: RPMI-1640, 10% FBS, 1% penicillin–streptomycin, and L-glutamine.
2. Collagenase type IV from *Clostridium histolyticum*.
3. DNase I.
4. Red blood Cell (RBC) lysis buffer.
5. FACS buffer: PBS, 0.1% BSA, 2 mM EDTA, and 0.05% sodium azide.
6. Freezing medium: 90% FBS–10% DMSO.
7. Freshly isolated liver cells or frozen-thawed liver cells. For sample type processing details, refer to Subheading 3.1.

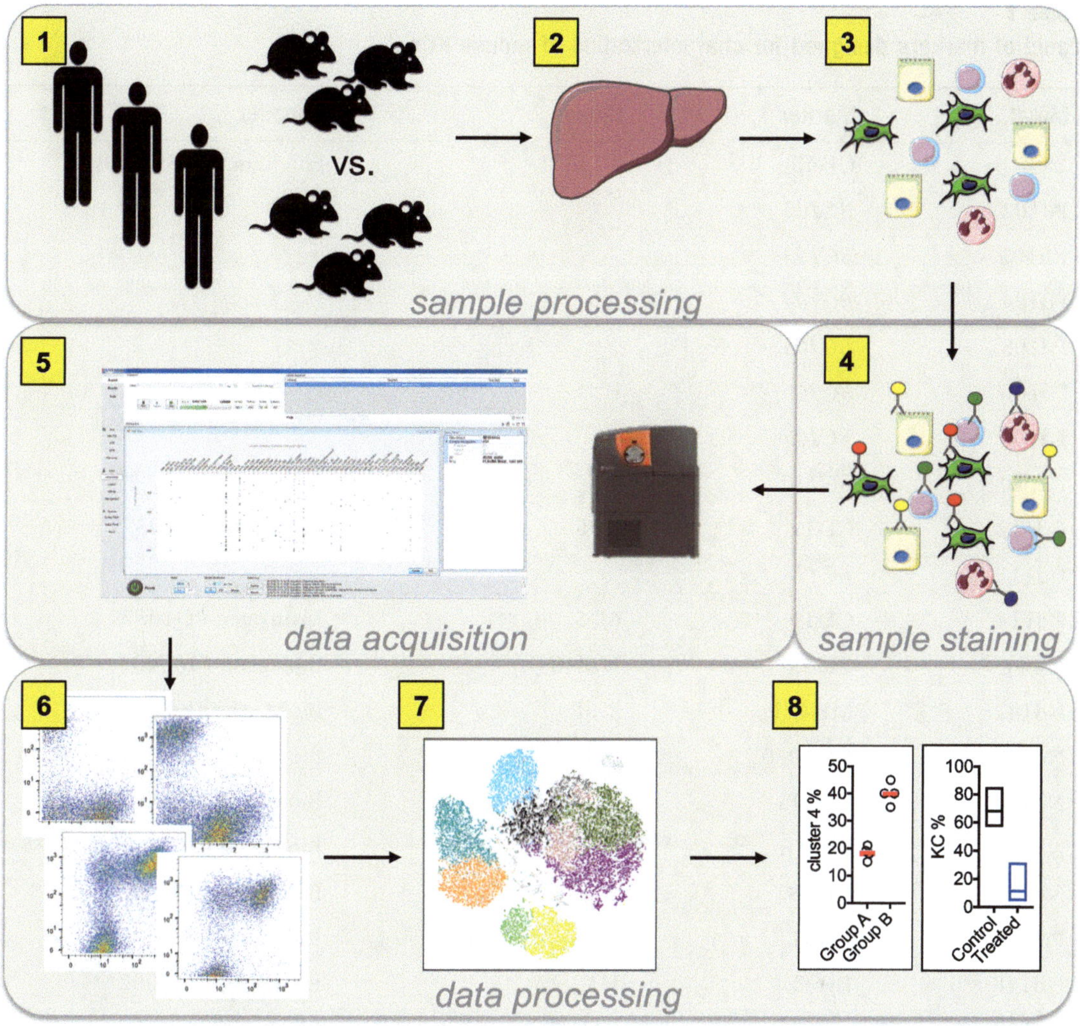


**Fig. 1** Workflow for Kupffer cell monitoring by CyTOF. (**1**) Establish the cohort of human patients enrolled in the study or design the mouse groups to analyze; (**2**) harvest the liver biopsies and (**3**) prepare the single-cell suspension; (**4**) process and label the samples and (**5**) acquire them on the CyTOF. (**6**) Postacquisition processing will generate .fcs files that can be then (**7**) manually and/or algorithm-based analyzed to generate final data

8. 0.4% trypan blue.
9. PBS buffer.
10. Double distilled water.
11. EQ Four Element Calibrations Beads (Fluidigm).
12. Metal-labeled antibodies (*see* panel Table 1).
13. Cell-ID cisplatin (Fluidigm).
14. Cell-ID intercalator–Ir (Fluidigm).
15. Mouse Fc Block.

**Table 1**
**Panel of markers designed for characterization of mouse KCs**

| Metal | Marker | Clone | Supplier |
|---|---|---|---|
| Y89 | CD45 | 30-F-11 | Fluidigm, #3089005B |
| Pd102 | *BC102* | | |
| Rh103 | *BC103* | | |
| Pd104 | *BC104* | | |
| Pd105 | *BC105* | | |
| Pd106 | *BC106* | | |
| Pd108 | *BC108* | | |
| Pd110 | *BC110* | | |
| Cd112 | CD19 | 6D5 | Invitrogen #Q10379 |
| Cd113 | *BC113* | | |
| Cd114 | CD19 | 6D5 | Invitrogen #Q10379 |
| Pr141 | CD48 | HM48-1 | Biolegend #103402 |
| Nd142 | MHCII | Y-3P | BioXCell #BE0178 |
| Nd143 | CD107b | M3/84 | Biolegend #108502 |
| Nd144 | CD11a | FD441.8 | BioXCell #BE005-1 |
| Nd145 | Ly6AE | D7 | Biolegend #108101 |
| Nd146 | CD88 | 20/70 | Biolegend #135802 |
| Pm147 | Ly6G | 1A8 | BD #551459 |
| Nd148 | Ly6C | HK1.4 | Biolegend #128002 |
| Sm149 | CD9 | KMC8 | BD #553758 |
| Sm150 | SiglecF | E50-2440 | BD #552125 |
| Eu151 | TIM4 | RMT4-54 | Biolegend #130004 |
| Sm152 | CD49b | DX5 | Biolegend #108902 |
| Eu153 | CD11b | M1/70 | BD #553308 |
| Sm154 | CD86 | GL1 | BD #553689 |
| Gd155 | CD68 | FA-11 | Bio-Rad #MCA1957 |
| Gd156 | BST2 | 120G8.04 | Novus Biologicals #DDX0390 |
| Gd157 | MerTK | Q60805 | R&Dsystems #BAF591 |
| Gd158 | CD11c | N418 | Biolegend #117302 |
| Tb159 | F4/80 | CI:A3-1 | Bio-Rad #MCA497 |
| Gd160 | CD206 | C068C2 | Biolegend #141702 |

(continued)

**Table 1**
**(continued)**

| Metal | Marker | Clone | Supplier |
|---|---|---|---|
| Dy161 | CD43 | S11 | Biolegend #143202 |
| Dy162 | CD81 | Eat2 | BD #559518 |
| Dy163 | CD226 | DNAM1-10E5 | eBioscience #16–2261-81 |
| Dy164 | CD103 | 2E7. | Biolegend #121402 |
| Ho165 | CD64 | X54–5/7.1 | Biolegend #139302 |
| Er166 | CD169 | 3D6.112 | Biolegend #142402 |
| Er167 | CCR2 | 475,301 | R&Dsystems #BAF5538 |
| Er168 | LYVE1 | ALY7 | eBioscience #14–0443 |
| Tm169 | CD102 | 3C4 | eBioscience #16–1021 |
| Yb170 | CD107a | 1D4B | Biolegend #121602 |
| Yb171 | SiglecH | 551 | Biolegend #129602 |
| Yb173 | CD172 | P84 | BD #552371 |
| Yb174 | CD24 | M1/69 | Biolegend #101802 |
| Lu175 | CD200R | OX-110 | Biolegend #123902 |
| Yb176 | CD90 | T24/31 | BioXCell #BE0212 |
| Ir191 | *DNA* | | |
| Ir193 | *DNA* | | |
| Pt195 | *Cisplatin* | | |

16. 16% PFA.
17. Bromoacetamidobenzyl-EDTA (BABE)-linked metal barcodes prepared in house as described in ref. [8].

### 2.2 Equipment and Supplies

1. Calibrated single/multichannel pipettes.
2. 5-ml microtubes.
3. 96-well U or V bottom plates.
4. Scissors and forceps.
5. 37 °C humidified incubator with 5% $CO_2$.
6. 10 ml-syringe equipped with an 18G needle.
7. 50-ml falcon tubes.
8. Cell strainers (70 μM).
9. Centrifuge with adaptors for 50-ml tubes.
10. 5-ml round-bottom polystyrene tubes with cell strainer caps.

11. 5-ml round-bottom polystyrene tubes.
12. Dual chamber counting slides (Bio-Rad).
13. Automated cell counter (Bio-Rad).
14. Freezing containers.
15. Helios, CyTOF system (Fluidigm).

## 3 Methods

### 3.1 Sample Processing

1. Prepare digestion buffer by adding collagenase (final concentration: 0.2 mg/ml) and DNase I (100 U/ml) to RPMI complete medium, 3 ml of digestion buffer are needed per sample (*see* **Note 1**).
2. Transfer 1 ml of digestion buffer in 5-ml Eppendorf tubes, 1 tube per sample.
3. Harvest liver biopsies and put them in the prefilled 5-ml Eppendorf tubes (*see* **Note 2**).
4. Cut the organs in very small pieces by using scissors (*see* **Note 3**).
5. Add 2 ml of digestion buffer.
6. Incubate the reaction mix at 37 °C for 30 min (*see* **Note 4**).
7. Homogenize cell suspensions by 5–10 passages through a syringe equipped with an 18G-needle (*see* **Note 5**).
8. Filter the solution on a cell strainer mounted on a 50-ml tube.
9. Rinse the filter with 10 ml of FACS buffer.
10. Centrifuge for 5 min at 1500 rpm (400 × *g*).
11. Aspirate the supernatant and wash the pellet with 10 ml of FACS buffer.
12. Centrifuge for 5 min at 1500 rpm (400 × *g*).
13. Discard the supernatant and lyse red blood cells by adding 500 μl RBC lysis buffer or ACK.
14. Transfer the solution to a 5-ml round-bottom polystyrene tube.
15. Incubate for 5 min at room temperature.
16. Add 3 ml of FACS buffer.
17. Centrifuge for 5 min at 1500 rpm (400 × *g*).
18. Aspirate the supernatant and wash the pellet with 3 ml of FACS buffer.
19. Centrifuge for 5 min at 1500 rpm (400 × *g*).

20. Resuspend the cells in 100 μl of FACS buffer (*see* **Notes 6** and 7).

From this step, cells are ready to be processed. If the experiment has to be delayed, cells can be frozen in freezing medium. In this case, cells have to be suspended in freezing medium instead of FACS buffer and stored at −80°.

### 3.2 Sample Staining

1. Transfer the cells into a 96-well U bottom plate (*see* **Note 8**).
2. Top up with PBS and centrifuge for 3 min at 1500 rpm (400 × *g*).
3. Resuspend the cells in 100 μl of FACS buffer and do viability staining using Cell-ID cisplatin. Usually, the viability is above 90%. If cells have been frozen and thawed, viability should be around 50%.
4. Incubate for 5 min on ice.
5. Top up with FACS buffer and centrifuge for 3 min at 1500 rpm (400 × *g*).
6. Resuspend in 100 μl of FACS buffer + Fc Block (1:200) to get rid of the nonspecific binding.
7. Incubate for 10 min on ice.
8. Top up with FACS buffer and centrifuge for 3 min at 1500 rpm (400 × *g*).
9. Resuspend the cells in 50 μl of the antibody cocktail prepared freshly according to their concentrations of use (*see* panel Table 1).
10. Incubate for 20 min on ice.
11. Top up with PBS and wash three times.
12. Fix the cells using 2% of PFA freshly prepared from 16% in 100 μl and leave the samples overnight at 4 °C (*see* **Note 9**). If the samples are also analyzed for intracytoplasmic and/or intranuclear markers, process to fixation/permeabilization and intracellular staining at this step according to the manufacturer's protocol. At the end of the intracellular staining, leave the samples in 100 μl of 2% PFA as previously stated.
13. The next day, spin down the cells for 3 min at 1500 rpm (400 × *g*), wash with PBS and spin down for another 3 min at 1500 rpm (400 × *g*).
14. Resuspend each sample in 100 μl of a specific barcode solution (*see* **Notes 10** and **11**). Mix well.
15. Incubate on ice for 30 min.
16. Top up with FACS buffer and centrifuge for 3 min at 1500 rpm (400 × *g*).

17. Stain DNA by resuspending the samples in 100 μl of Cell-ID intercalator–Ir (1:2000 dilution from stock in FACS buffer).
18. Incubate for 10 min on ice.
19. Wash with FACS buffer and centrifuge for 3 min at 1500 rpm (400 × *g*).
20. Resuspend in 100 μl of FACS buffer.
21. Transfer and filter the samples on a 5-ml round-bottom polystyrene tube with cell strainer cap. If the samples are barcoded, they can be combined into the same tube. Wash three times with double distilled water (*see* **Note 12**).
22. Count the cells and adjust the concentration to $5 \times 10^5$ cells $ml^{-1}$ in double distilled water (*see* **Note 13**).
23. Add 1–2% of EQ Four Element Calibrations Beads (Fluidigm) and mix well.
24. Run the samples on the mass cytometer.

### 3.3 Data Acquisition

Start and set up the machine as described previously [9]. The operation and maintenance of the CyTOF instrument requires specifically qualified personnel. Briefly, in the context of a daily operation, it requires the following steps:

1. Preparation and startup of the instrument including connecting the nebulizer to the CyTOF, checking the nebulizer spray, and warming up of the instrument followed by the plasma startup.
2. Two washing steps with the washing buffer first and then double distilled water.
3. Daily quality control that includes checking the instrument performance, monitoring the background and tuning the system using the CyTOF tuning solution. To have the tuning procedure to pass successfully, the following parameters are essential: resolution mass, mean duals, dual slopes, R2, oxide ratio, and relative standard deviation (%RSD) (For additional information, check the Helios User Guide available online).
4. Two additional washing steps with the washing buffer first and then double distilled water, until the instrument is clean.
5. Data acquisition. Set up the panel in the Experiment Manager window, load the sample onto the sample loader, preview to check the quality of the sample and record data. At the end of each acquisition, one .fcs file will be generated.

### 3.4 Data Analysis

The generated .fcs files can be manipulated exactly as the ones resulting from conventional flow cytometry even if a compensation matrix is not required for mass cytometry data. A classical analysis can be done by using FlowJo software or equivalent. In this

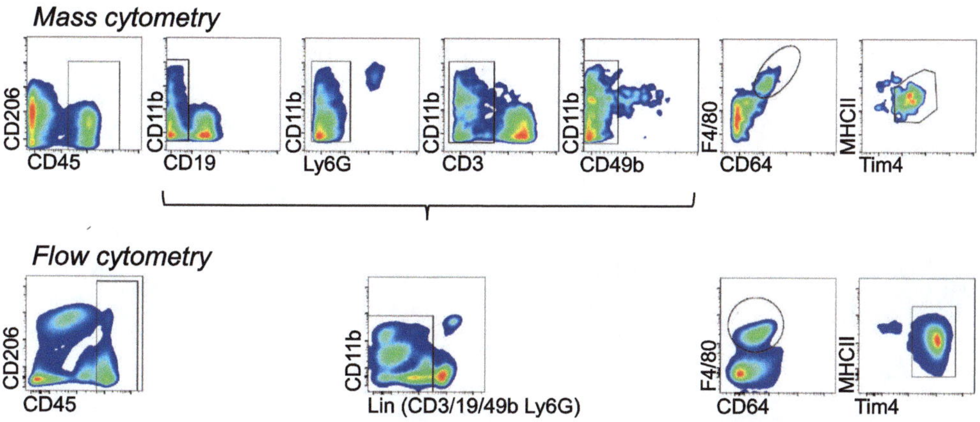


**Fig. 2** Comparison between flow and mass cytometries for mouse KC gating. Liver cells were prepared as described and analyzed by mass cytometry (Helios, Fluidigm) or classical flow cytometry (LSRII 5 lasers, BD). KCs are defined by conventional flow cytometry as CD45$^+$ Lin$^-$ CD11b$^{int}$ CD64$^+$ F4/80$^+$ MHCII$^{int+}$ Tim4$^+$ cells. Mass cytometry allows for the resolution of the lineage and offers a larger number of free channels to analyze marker expression. In this example, eleven markers are needed to define KCs: CD3, CD11b, CD19, CD45, CD49b, CD64, CD206, F4/80, Ly6G, MHCII, and Tim4. Due to fluorescence spectrum spillovers, the use of additional markers of interest to better characterize KCs is strongly limited by using classical flow cytometry. Using mass cytometry that offers 30 other free channels circumvents these limitations

context, a manual gating strategy can be applied to look at the expression on several markers in different subpopulations. As shown in Fig. 2, both mass and flow cytometries allow for an efficient isolation of the CD45$^+$ Lin$^-$ CD11b$^{int}$ CD64$^+$ F4/80$^+$ MHCII$^{int}$ Tim4$^+$ KCs. But the main benefit of using mass cytometry is the large panel of markers that can be detected simultaneously. Due to the inherent complexity of these large panels, it can be meaningless to manually define the populations by using hierarchical gating on two-dimensional dot plots. Automated algorithm-based clustering represents a more efficient strategy to analyze data as they allow for the reduction of dimensions, integrating the expression of all of the markers and offering a two-dimensional visualization of the dataset. FlowJo software now integrates plugins allowing for the one-click generation of t-distributed stochastic neighbor embedding (t-SNE) [10] and/or Uniform Manifold Approximation and Projection (UMAP) visualizations [11]. An efficient, user-friendly, and more powerful alternative is the use of the cytofkit R package [12]. As an illustrating example, Fig. 3 shows data from a C57/BL6 mouse liver stained with a 36-marker panel as analyzed by cytofkit. The population of CD45$^+$ CD11b$^{int}$ CD64$^+$ F4/80$^+$ MHCII$^{int}$ Tim4$^+$ KCs is easily identifiable, but mass cytometry also reveals the expression of other markers such as CD81, CD107a, CD107b, CD169, CD206, and MerTK. Following the expression of these markers in specific conditions

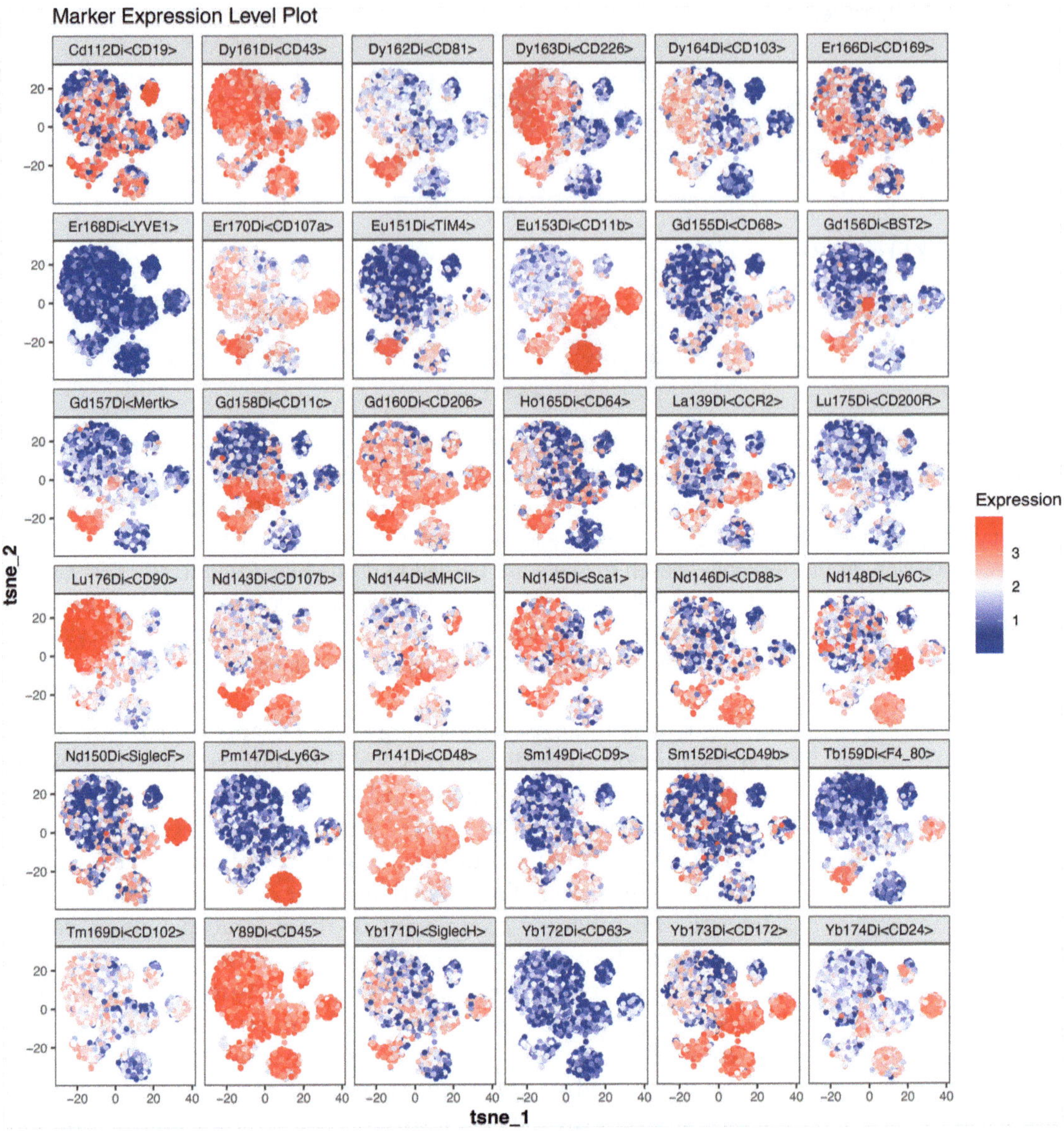


**Fig. 3** Liver immune cell characterization by mass cytometry. This analysis was done by using the cytofkit R package. Briefly, the sample was processed, stained and acquired as described. CD45$^+$ immune cells were manually gated and an individual .fcs was generated. A t-SNE visualization was generated by integrating the expression of 36 markers and the level of expression of each marker is projected on it

(diseases, diets, infection, ...) could provide meaningful information on the biology of KCs. But whatever the strategy that will be used, the analysis should be driven by keeping in mind the initial question as mass cytometry data come with a lot of dimensions. So it is really important to define clearly the question to reduce time and cost of the analysis.

## 4 Notes

1. Collagenase is thermosensitive and cannot be conserved in working solutions more than few hours. Furthermore, it should not be frozen/thawed more than one time. For convenience, after resuspension of the powder at a concentration of 10 mg $ml^{-1}$, 1-ml aliquots should be frozen at −20°. For each experiment, just thaw an aliquot and dilute it in 50 ml of RPMI complete medium. The same strategy can be used for DNase I. Discard the remaining digestion buffer that might not have been used at the end of the experiment.
2. The liver tissue contains a lot of lipids released during the digestion that can destabilize the pellet after centrifugation. To circumvent this, only one lobe of the liver can be processed. It should provide enough cells for a standard analysis. If a lot of cells are required, it is better to separate the whole liver in different tubes to avoid the loss of cells during supernatant aspirations in the subsequent steps.
3. For an efficient digestion, the liver has to be cut in very small pieces before incubation in digestion buffer. One efficient method is to put the sample in 500 μl of digestion buffer in a 2-ml Eppendorf tube and then, directly cut it by rapid scissor movements inside the tube.
4. The liver is a soft organ that can be easily dissociated even without enzymatic digestion, notably when sensitive markers have to be detected (such as CD11c or CD115). In this case, the organ can be mechanically dissociated by putting it on a cell strainer and crushing it with a syringe plunger.
5. The passages through the needle after incubation are really important for an efficient dissociation. So this step should not be overlooked. After this, the solution should be almost homogeneous with no remaining macroscopic pieces. If the needle clogs repeatedly by undigested pieces, it could be necessary to extend the incubation time.
6. Numerous protocols of liver cell preparations include an enrichment of immune cells by a Percoll gradient (1 or 2 layers depending on the protocols). Although it effectively enriches the sample in immune cells, it may result in the selective loss of subpopulations and notably KC. In order to have the best representation of the diversity of the cells, it is better to avoid the Percoll-mediated purification.
7. A similar remark can be made about the magnetic enrichment. Common magnetic beads used to purify macrophages use CD45 and CD11b antibodies but as KC express these two markers at a quite low level, a purification strategy using such beads may results in the loss of cells.

8. If the number of cells in the samples is low, the 96-well V bottom plate could be a good option (reduced loss of cells by centrifugation and easier to distinguish the pellet).
9. From there, the samples are usually run on the CyTOF the following day. The cells can stay in the fridge for up to 4 days until the run. If the experiment has to be delayed, samples have to be frozen down at −80 °C in freezing medium, the latest 4 days after staining. In that case, cells are then washed off the PFA on D2, resuspended in 90%FBS + 10%DMSO and stored at −80 °C.
10. Cell barcoding allows for simultaneous acquisition of different samples and helps to reduce the intersample variations. Our protocol describes a fixed cell barcoding system but other techniques using CD45 [13, 14] or beta-2-microglobulin + ATPase-subunit CD298 [15] for live cell barcoding have been reported as well.
11. The barcode solution depends on the concentration of use of each barcode and the number of barcodes assigned for each sample. As an example, we have a batch of barcodes Pd102 and Pd104 used at 1:4000 final, Pd106 used at 1:1000 final. For a double-barcoded sample using Pd102 and Pd104, we will prepare a 1:2000 dilution of both barcodes, and add 50 μl of each to the sample. For a triple-barcoded sample using Pd102, Pd104, and Pd106, a 1:1333 dilution will be prepared for Pd102 and Pd104, a 1:333 dilution will be prepared for Pd106, and 33 μl of each will be added to the sample.
12. It is crucial to wash the samples at least three times to minimize buildups of salts in the machine.
13. The length of a run depends on the volume of the sample. We recommend not exceeding the concentration of the cells above $5 \times 10^5$ cells $ml^{-1}$, which is equivalent to an acquisition rate of ~200–300 cells $s^{-1}$. This will significantly avoid getting doublets and backgrounds and allow for data of much higher quality. As a reference, The Helios CyTOF takes a volume of 30 μl $min^{-1}$, which is roughly 1 h to acquire $1 \times 10^6$ cells in 2 ml of double distilled water.

## References

1. Bandura DR, Baranov VI, Ornatsky OI, Antonov A, Kinach R, Lou X, Pavlov S, Vorobiev S, Dick JE, Tanner SD (2009) Mass cytometry: technique for real time single cell multitarget immunoassay based on inductively coupled plasma time-of-flight mass spectrometry. Anal Chem 81(16):6813–6822. https://doi.org/10.1021/ac901049w
2. Lavin Y, Kobayashi S, Leader A, Amir ED, Elefant N, Bigenwald C, Remark R, Sweeney R, Becker CD, Levine JH, Meinhof K, Chow A, Kim-Shulze S, Wolf A, Medaglia C, Li H, Rytlewski JA, Emerson RO, Solovyov A, Greenbaum BD, Sanders C, Vignali M, Beasley MB, Flores R, Gnjatic S, Pe'er D, Rahman A, Amit I, Merad M (2017) Innate immune landscape in early lung adenocarcinoma by paired single-cell analyses. Cell 169(4):750–765.e717. https://doi.org/10.1016/j.cell.2017.04.014

3. Simoni Y, Fehlings M, Kloverpris HN, McGovern N, Koo SL, Loh CY, Lim S, Kurioka A, Fergusson JR, Tang CL, Kam MH, Dennis K, Lim TKH, Fui ACY, Hoong CW, Chan JKY, Curotto de Lafaille M, Narayanan S, Baig S, Shabeer M, Toh SES, Tan HKK, Anicete R, Tan EH, Takano A, Klenerman P, Leslie A, Tan DSW, Tan IB, Ginhoux F, Newell EW (2017) Human innate lymphoid cell subsets possess tissue-type based heterogeneity in phenotype and frequency. Immunity 46(1):148–161. https://doi.org/10.1016/j.immuni.2016.11.005
4. Guilliams M, Scott CL (2017) Does niche competition determine the origin of tissue-resident macrophages? Nat Rev Immunol 17 (7):451–460. https://doi.org/10.1038/nri.2017.42
5. Bleriot C, Ginhoux F (2019) Understanding the heterogeneity of resident liver macrophages. Front Immunol 10:2694. https://doi.org/10.3389/fimmu.2019.02694
6. Sanyal A, Boyer T, Terrault N, Lindor K (2017) Zakim and Boyer's hepatology. A textbook of liver disease, 7th edn. Elsevier
7. Halpern KB, Shenhav R, Matcovitch-Natan O, Toth B, Lemze D, Golan M, Massasa EE, Baydatch S, Landen S, Moor AE, Brandis A, Giladi A, Avihail AS, David E, Amit I, Itzkovitz S (2017) Single-cell spatial reconstruction reveals global division of labour in the mammalian liver. Nature 542(7641):352–356. https://doi.org/10.1038/nature21065
8. Sumatoh HR, Teng KW, Cheng Y, Newell EW (2017) Optimization of mass cytometry sample cryopreservation after staining. Cytometry A 91(1):48–61. https://doi.org/10.1002/cyto.a.23014
9. Leipold MD, Maecker HT (2012) Mass cytometry: protocol for daily tuning and running cell samples on a CyTOF mass cytometer. J Vis Exp 69:e4398. https://doi.org/10.3791/4398
10. VanderMaaten L, Hinton G (2008) Visualizing data using t-SNE. J Mach Learn Res 9:2579–2605
11. Becht E, McInnes L, Healy J, Dutertre CA, Kwok IWH, Ng LG, Ginhoux F, Newell EW (2019) Dimensionality reduction for visualizing single-cell data using UMAP. Nat Biotechnol. https://doi.org/10.1038/nbt.4314
12. Chen H, Lau MC, Wong MT, Newell EW, Poidinger M, Chen J (2016) Cytofkit: a bioconductor package for an integrated mass cytometry data analysis pipeline. PLoS Comput Biol 12(9):e1005112. https://doi.org/10.1371/journal.pcbi.1005112
13. Han G, Spitzer MH, Bendall SC, Fantl WJ, Nolan GP (2018) Metal-isotope-tagged monoclonal antibodies for high-dimensional mass cytometry. Nat Protoc 13 (10):2121–2148. https://doi.org/10.1038/s41596-018-0016-7
14. Lai L, Ong R, Li J, Albani S (2015) A CD45-based barcoding approach to multiplex mass-cytometry (CyTOF). Cytometry A 87 (4):369–374. https://doi.org/10.1002/cyto.a.22640
15. Hartmann FJ, Simonds EF, Bendall SC (2018) A universal live cell barcoding-platform for multiplexed human single cell analysis. Sci Rep 8(1):10770. https://doi.org/10.1038/s41598-018-28791-2

# Chapter 11

# Enzymatic Quantification of Liver Lipids After Folch Extraction

Mirko E. Minniti, Osman Ahmed, and Matteo Pedrelli

## Abstract

The determination of the lipid content in liver offers the potential to investigate metabolic alterations in different research contexts. Here, we describe a method to determine cholesterol, triglycerides, and phospholipids in liver samples based on total lipid isolation by a 2:1 chloroform–methanol mixture (Folch extraction) and specific enzymatic colorimetric microassays in plate.

**Key words** Liver, Cholesterol, Triglycerides, Phospholipids, Enzymatic assay

## 1 Introduction

Liver is the major organ regulating the metabolism of lipids, such as cholesterol and triglycerides, and perturbations of its homeostasis can affect the amount and composition of fat stored within. Therefore, the quantification of the lipid content can be a useful tool to explore several metabolic pathways. To determine the lipids from tissues, it is crucial to properly isolate the nonpolar species of interest and to remove potential polar contaminants. This extraction step can be performed with different solvents (alone or in combination) depending on the properties of both the tissue matrix and the lipids intended to analyze. The classic method to isolate total lipids from animal tissues is known as "Folch extraction." This protocol was published by Jordi Folch and colleagues in the 1950s [1], and consisted in using 20 parts of a chloroform–methanol mixture 2:1 by volume per 1 part of sample. Alternatively, another extraction procedure commonly used to isolate total lipids from water-rich tissues is the Bligh–Dyer method, which applies 3 parts of 1:2 chloroform–methanol per 1 part of sample followed by 1 or 2 parts of chloroform [2].

Myriam Aouadi and Valerio Azzimato (eds.), *Kupffer Cells: Methods and Protocols*, Methods in Molecular Biology, vol. 2164,
https://doi.org/10.1007/978-1-0716-0704-6_11, 

When these protocols were established, the quantification of lipids from the extracts mainly relied on time-consuming chromatographic and chemical analyses for every individual lipid. This technique was improved and scaled down in the 1990s by the group of Lawrence L. Rudel [3], which combined a simplified version of the Folch extraction with colorimetric enzymatic assays that were available for serum lipids. Here, we describe an optimized method routinely run in our laboratory which comprises: (a) the extraction of the lipids from a liver sample, (b) the colorimetric enzymatic quantification of total and free (or unesterified) cholesterol, triglycerides and phospholipids from the same extract, and (c) the determination of the protein content from the sample, useful to correct the lipid quantity for the initial amount of tissue. Small modifications of this protocol can be applied to extract lipids from different biological tissues (*e.g.*, aorta or muscle biopsies), and even cultured cells.

## 2 Materials

To ensure reproducible results, use always ultrapure water (type 1) and analytical-grade reagents. When handling hazardous reagents use glassware and chloroform-compatible plastics (*e.g.*, polytetrafluoroethylene/PTFE), and work carefully in a fume hood. Follow the local regulations when disposing waste materials.

### 2.1 Folch Extraction

1. 50–100 mg of snap-frozen liver.
2. Glass tubes for evaporation with screw caps in PTFE.
3. Folch solution: mix 2 volumes of chloroform with 1 of methanol in a brown glass container (*see* **Note 1**). Store at room temperature in a fireproof cabinet.
4. 0.05% $H_2SO_4$.
5. Centrifuge for glass tubes.
6. Dry block incubator for glass tubes with needles connected to nitrogen gas cylinder.
7. Chloroform.
8. 1% (*v*/*v*) Triton X-100 in chloroform (*see* **Note 2**). Store at room temperature in a fireproof cabinet.

### 2.2 Lipid Quantification

We use commercially available colorimetric enzymatic in vitro tests for the quantitative determination of the lipids that can be performed in 96-well plates. For every lipid (total and free cholesterol, triglycerides, and phospholipids) there are a specific reagent and a standard to be used (free cholesterol standard is used for both total and free cholesterol).

1. Total cholesterol: the assay for total cholesterol is based on the cholesterol oxidase and the chromogenic indicator system phenol/4-aminophenazone (CHOD-PAP method), after the enzymatic hydrolysis of cholesteryl esters [4]. CHOL2 (Roche Diagnostics) has a measuring range of 0.1–20.7 mM (3.86–800 mg/dL). To generate the standard curve, we use the standard solution of free cholesterol (see below). Alternatively, for both total and free cholesterol a more sensitive fluorometric assay can be used, such as the Amplex™ Red Cholesterol Assay Kit (Thermo Fisher Scientific).
2. Free cholesterol: the assay for free cholesterol is similar to the total cholesterol one, but without the step of cholesteryl ester cleavage. Free Cholesterol E (FUJIFILM Wako Chemicals) includes the free cholesterol standard.
3. Triglycerides: this assay is based on the glycerol-3-phosphate oxidase and the chromogenic substrates 4-chlorophenol/4-aminophenazone (GPO-PAP method) [5]. TRIGL (Roche Diagnostics) has a measuring range of 0.1–10.0 mM (8.85–885 mg/dL). As standard, we use a control based on human serum with a known concentration of triglycerides, usually in the normal range (Precinorm U, Roche Diagnostics).
4. Phospholipids: the assay to measure total phospholipids (lecithin, lysolecithin, sphingomyelin) is based on the choline oxidase and a chromogenic indicator system, following the enzymatic hydrolysis of choline-containing phospholipids [6]. LabAssay™ Phospholipid (FUJIFILM Wako Chemicals) is linear up to 600 mg/dL and includes a standard solution of phospholipids.
5. Spectrophotometer or plate reader.

#### 2.3 Protein Determination

1. 0.25 M NaOH.
2. Modified Lowry microassay in plate (or similar protein assays compatible with NaOH): *DC*™ Protein Assay (Bio-Rad) has a protein concentration range of 5–250 μg/mL and is compatible with 0.5 M NaOH.
3. Spectrophotometer or plate reader.

## 3 Methods

Carry out all procedures at room temperature, unless otherwise specified. Remove chloroform from all glassware and evaporate the leftover in the fume hood before cleaning or disposing of.

#### 3.1 Folch Extraction

1. Put the liver piece (50–100 mg) into a glass tube.
2. Add 5 mL of Folch solution per glass tube (*see* **Note 3**).

3. Cap and gently vortex the glass tubes till the tissue floats within the solution.
4. Incubate the glass tubes at 4 °C for 12–72 h (*see* **Note 4**).
5. Pipet the Folch solution in a new tube and keep it capped.
6. Dry the livers in the tubes from **step 4** at 60 °C. Save the tubes for the protein determination (Subheading 3.3) (*see* **Note 5**).
7. In the meantime, add 1 mL of 0.05% $H_2SO_4$ to the Folch solution from **step 5** (*see* **Note 6**).
8. Cap and gently vortex the glass tubes to mix the phases thoroughly.
9. Centrifuge the glass tubes at 500 × *g* for 15 min to separate the phases.
10. Discard the upper aqueous phase and pipet the lower organic phase into a new glass tube.
11. Dry the organic phase at 60 °C in the dry block incubator under nitrogen atmosphere (*see* **Note 7**).
12. Add to each tube 5 mL of chloroform, cap and gently vortex (*see* **Note 8**).
13. Pipet 0.5 mL of the chloroform solution from **step 12** into a new glass tube (*see* **Note 9**).
14. Add 1 mL of 1% Triton X-100 in chloroform (*see* **Note 10**).
15. Dry completely the chloroform at 40–45 °C in the dry block incubator under nitrogen atmosphere (*see* **Note 11**).
16. Add 0.5 mL of water.
17. Incubate for 5 min at 60 °C to solubilize the Triton X-100 and the lipids in water.
18. Vortex until the solution becomes completely clear and transparent (*see* **Note 12**).

### 3.2 Lipid Quantification

1. Pipet each sample (including the blank) from Subheading 3.1, **step 18** in duplicate in a 96-well plate, together with the standard points for the appropriate lipid (*see* **Note 13**).
2. Add the total cholesterol, free cholesterol, triglyceride, or phospholipid reagent for each sample or standard point (*see* **Note 14**).
3. Incubate the plate for 20–30 min at room temperature or 10 min at 37 °C, according to the manufacturer's instructions.
4. Read the plate at the specific absorbance depending on the lipid reagent.
5. Calculate the lipid mass by interpolation of the absorbance on the standard curve.
6. Correct for the dilution factor to retrieve the lipid mass in the liver sample extracted (*see* **Note 15**).

### 3.3 Protein Determination

1. Add 1 mL of 0.25 M NaOH to the tubes from Subheading 3.1, **step 6** (*see* **Note 16**).
2. Cap and heat the tubes at 60 °C until the tissue is completely dissolved and the protein digested (*see* **Note 17**).
3. Determine the correct protein dilution to be measured (*see* **Note 18**).
4. Pipet each sample (including the blank) in triplicate in a 96-well plate, together with the standard points for the total protein.
5. Add the microassay reagents, incubate the plate and measure the absorbance based on the assay.
6. Calculate the protein mass by interpolation of the absorbance on the standard curve.
7. Correct for the dilution factor to retrieve the protein mass in the liver sample (*see* **Note 19**).

## 4 Notes

1. Prepare the Folch solution at least 12 h before use. Invert several times the glass container to allow proper mixing in one single phase (being careful in handling). Open the lid to release the gas pressure after turning the container.
2. Triton X-100 is very viscous to pipet; we cut off the plastic tip or use the reverse pipetting technique.
3. The volume of Folch solution per part of sample used in the original method is 20:1 (or 20 mL/g of sample) [1]. In this protocol, we use a larger excess of Folch solution (50–100:1), both to improve the extraction and to avoid using too small volumes of organic solvents. It is important not to use a smaller proportion compared to the original method.
4. Incubate the tubes in a fireproof fridge. Alternatively, the extraction can be performed at room temperature (for 6–72 h, depending on the lipid content) or speeded up at 60 °C for 2 h (in a fume hood). Based on our experience, lipids are completely extracted when the tissue falls at the bottom of the tube.
5. To our experience, the evaporation takes between 30 min and 1 h. Tubes can be stored at −20 °C until protein digestion and determination.
6. Based on the protocol from Rudel's group, the volume of 0.05% $H_2SO_4$ is 1/5 of the Folch solution used [3].

7. Regulate gently the intensity of the nitrogen flow through the needles not to splash the glass tube walls. To our experience, the evaporation takes between 30 min and 2 h depending on the nitrogen flow.
8. Because Folch solution is very volatile, this resolubilization is extremely important to avoid volume losses that could have occurred during the previous steps. The volume of chloroform used is not fixed, and can be modified depending on the expected amount of lipids within the sample.
9. The remaining 4.5-mL lipid extract in chloroform can be capped and stored for several months at −20 °C, 4 °C or room temperature. Mark the volume on the tubes to resolubilize the lipid extract in the same volume of chloroform in case of evaporation. Again, the volume of the chloroform aliquot is not fixed and can be modified depending on the expected amount of lipids within the sample. In case the analysis had to be repeated because of excessive lipid dilution, it is possible to take a larger aliquot or to reconcentrate the whole extract (being careful in keeping track of the dilution factors).
10. It is good practice to add an extra tube with only 1% Triton X-100 in chloroform to be used as a sample blank in the subsequent lipid quantifications. Moreover, the standard solutions of the different lipids should be prepared in chloroform and treated as the samples from this point onward, as explained [3]. The use of the commercially available assay-included standards is allowed, although we suggest to validate their suitability during the technique standardization.
11. Triton X-100 is used to solubilize in water the lipids from the extract in chloroform. It is thus crucial to remove completely the chloroform in this step. This can be checked by tilting the tube: when only a dense drop of Triton X-100 is present in the tube, the chloroform is evaporated. To our experience, this evaporation can take up to 2 h (depending on the nitrogen flow).
12. A common problem in this step is the water solution appearing cloudy. This could be due to the presence of chloroform, hence requiring some cycles of heating at 60 °C and vortexing to be removed. If not enough to attain a clear solution, the cause could rely on the excessive amount of triglycerides in the sample that cannot be solubilized with the Triton X-100 in water. In this case, it is necessary to take a smaller aliquot from Subheading 3.1, **step 12** (or to reprepare the sample) and/or to use a higher volume of 1% Triton X-100 in chloroform in Subheading 3.1, **step 14**, adjusting the water volume accordingly. To prevent interferences of the detergent with the

subsequent enzymatic reactions, the final concentration of Triton X-100 in the lipid quantification assay should not be higher than 0.5% [3]. The water solution can be transferred in a 1.5-mL tube and stored at 4 °C (or −20 °C for longer storage).

13. Use separate plates for the different lipid determinations. With this liver mass and a standard curve of 1–10 μg/well for total and free cholesterol, use 50 μL of sample in water solution. According to our experience, 10–50 μL sample for triglycerides are enough with a standard curve of 2–25 μg/well. These numbers are just an indication and may be modified depending on the samples analyzed.
14. The reagent volumes depend on the assay specifications. To our experience, the use of 150 μL reagent is fine with all the suggested assays and sample volumes.
15. The dilution factor for 50 μL out of 500 μL water solution from 0.5 mL of the original 5 mL Folch extract is 100. It is also possible to calculate the amount of cholesteryl esters in the sample by subtracting the moles of free cholesterol from those of total cholesterol. If the mass is used instead, after subtracting the free cholesterol from the total cholesterol, multiply the result by 1.67. This conversion factor takes into account the higher molar mass of cholesteryl esters (625–651 g/mol, depending on the ester) compared with unesterified cholesterol (387 g/mol).
16. As described for the lipid quantification after Folch extraction, it is good practice to add an extra tube with only 0.25 M NaOH to be used as a sample blank in the protein determination. Likewise, the standard solution for total protein (*e.g.*, bovine serum albumin) should be prepared in NaOH and treated as the samples from this point onward.
17. To our experience, the protein digestion takes between 30 min and 2 h depending on the samples. Vortex the samples from time to time to speed up the digestion. The water solution can be transferred in a 1.5-mL tube and stored at 4 °C (or −20 °C for longer storage).
18. With a standard curve of 0.2–1.4 μg/μL protein, a dilution 1:50 should be fine.
19. Use the total protein amount to correct the lipid quantity for the initial amount of tissue.

## References


1. Folch J, Lees M, Sloane Stanley GH (1957) A simple method for the isolation and purification of total lipides from animal tissues. J Biol Chem 226(1):497–509
2. Bligh EG, Dyer WJ (1959) A rapid method of total lipid extraction and purification. Can J Biochem Physiol 37(8):911–917. https://doi.org/10.1139/o59-099
3. Carr TP, Andresen CJ, Rudel LL (1993) Enzymatic determination of triglyceride, free cholesterol, and total cholesterol in tissue lipid extracts. Clin Biochem 26(1):39–42
4. Allain CC, Poon LS, Chan CS, Richmond W, Fu PC (1974) Enzymatic determination of total serum cholesterol. Clin Chem 20(4):470–475
5. Siedel J, Schmuck R, Staepels J, Town MH (1993) Long term stable liquid ready-to-use monoreagent for the enzymatic assay of serum or plasma triglycerides (GPO-PAP method). Clin Chem 39(6):1127
6. Takayama M, Itoh S, Nagasaki T, Tanimizu I (1977) A new enzymatic method for determination of serum choline-containing phospholipids. Clin Chim Acta 79(1):93–98

# Chapter 12

# In Vivo Investigation of High-Fat Diet–Induced Hepatic Lipid Dysfunctions

**Marion Korach-André**

## Abstract

Fat distribution, on top of general obesity, contributes to the severity of histologic features in patients with nonalcoholic fatty liver diseases (NAFLD); and visceral obesity has been correlated to fatty liver diseases. Therefore, investigation of fat distribution in vivo could be a good predictor of fatty liver risks in obesity. Fatty acids composition is a key player in hepatic dysfunctions and cardiovascular risk in obesity. Because fatty acids can damage biological membranes, fatty acid accumulation in the liver may be partially responsible for the functional and morphological changes that are observed in NAFLD. Fatty acids stored into triglycerides are lipid species that act as signaling molecules and therefore are key regulators of posttranslational regulation of biological functions such as lipid homeostasis and lipotoxicity. Here, we describe magnetic resonance methods to investigate in vivo whole-body fat distribution and hepatic liver fatty acid composition in order to directly assess the liver metabolic status and may allow to anticipate liver diseases.

**Key words** Nuclear magnetic resonance, Magnetic resonance imaging, Fat distribution, Magnetic resonance spectroscopy, In vivo, Lipids, Liver steatosis, Obesity

## 1 Introduction

Rodent models are commonly used in medical research to investigate a large spectrum of diseases. Selecting the appropriate animal model as model of human disease is a key aspect in outstanding medical research. The significance of results from animal experiments is therefore dependent of a suitable animal model. In this chapter, we will focus on the study of biological processes to gain insights into the physiological and pathophysiological response in obesity and associated liver dysfunctions, including NAFLD. When performing laboratory animal experiments, metabolic disorders investigated in animal models should be possible to infer the outcomes and extrapolate them to humans. This implies that the model studied in which a spontaneous or induced pathological process is investigated resembles, at least in one aspect, humans. For example, rare mutations in human pathology exist but do not

Myriam Aouadi and Valerio Azzimato (eds.), *Kupffer Cells: Methods and Protocols*, Methods in Molecular Biology, vol. 2164,
https://doi.org/10.1007/978-1-0716-0704-6_12, 

represent a large number of the human population. For this reason, knockout animal models should be carefully used to understand in vivo metabolic adaptations in human diseases. Consequently, when embarking upon the study of specific metabolic characteristics, it is essential to select appropriate species or strain.

In addition to the strain, a particular attention should be given to the diets used as a model for diet-induced obesity (DIO) including the control diet. Indeed, a large number of metabolic studies use high-fat diet (HFD) containing a large amount of fat coming from sources that can barely be comparable to human diet. However, a clear relationship between obesity and dietary fat content has been observed in large human epidemiological studies [1, 2], as well as in rodents [3]. Therefore, the use of dietary manipulation to promote obesity and associated metabolic dysfunction in animal models is relevant. Nevertheless, the amount of fat [4] and the fatty acid composition of the diet can drastically modify the metabolic adaptation to the HFD [5–7]; hence, lipid composition into the diet must be carefully monitored and might be a key factor in the discrepancy observed in the literature when it comes to DIO mouse model experiments. Finally, the literature recently demonstrated that the sex used in metabolic studies is a key element in the metabolic outcomes in DIO models both in rodents [8] and humans [5]. In brief, sex matters and should definitely be clearly define in any metabolic experimental protocol.

Overeating behavior associated with reduced energy expenditure is one of the major threats for obesity and associated metabolic dysfunctions, nowadays, in humans. However, this process is slow, so to better mimic human disorders associated with obesity, fat content in animal models should be tightly adjusted.

## 2 Materials

### 2.1 Animal Models

In obesity related metabolic studies, the most common mouse strain used is the so-called wild-type C57Bl/6J mouse. The C57Bl/6J mouse substrain is highly responsive to HFD-induced obesity and presents several features observed in human obesity including liver steatosis and insulin resistance [9]. It is also the background strain for the commonly used *ob/ob* and *db/db* mouse [10].

1. C57/Bl6J mouse substrain for high-fat diet–induced obesity mouse model.
2. ob/ob mouse strain for overeating obesity mouse model.
3. db/db mouse strain for high-fat diet–induced obesity and type-2 diabetes mouse model (*see* **Note 1**).

### 2.2 Diet

In obesity related metabolic studies, the most common model used is the high-fat diet–induced obesity model. There are several providers of high-fat diets for rodent studies. As mentioned above, the choice of the diet should be carefully set in terms of fat content and lipid composition. In addition, the match control diet used should be set for each high fat diet to be able to compare the effect of the increase fat content solely. When changing providers, you may see changes in the metabolic response to the high-fat diet; every formula is different and may be differently metabolized by the body (*see* **Note 2**).

1. High-fat diet (Research Diet®, D12451, 45% Kcal from fat; made from soybean oil (12%, no phytoestrogens to minimize the activation of estrogen receptors by the diet) and lard (88%), is used in our study.
2. Match control diet (Research Diet®, D12450H, 10% Kcal from fat, made from soybean oil (55%, containing no phytoestrogens) and lard (45%).

### 2.3 MR Scanner, Coil, and Monitoring Equipment

1. Several companies which develop magnetic resonance (MR) scanners for both animal and human studies (*see* **Note 3**) exist. The MR scanner used in the current protocol is a Varian 9.4 T with a 31 cm bore size. The main companies are Bruker, Siemens, and Varian. They all provide similar functionalities and the choice of the company will depend on the agreement you can negotiate when you purchase the equipment.
2. Respiration, core body temperature, and electrocardiogram (ECG) are monitored using the Monitoring and Gating products (SA Instruments, Stony Brook, NY, USA. http://www.i4sa.com/web_app/main/defaultProduct.aspx?ID=15&PT=3).
3. As for the MR scanner, a large variety of coils can be used to optimize the resolution of the image and the spectrum obtained in magnetic resonance research (*see* **Note 4**). In addition, NMR centers may develop their own coil according to their research interest.

## 3 Methods

Animals are anesthetized using isoflurane (4% for induction of deep sleep and ~2% maintenance) in a 3:7 mixture of oxygen and air, before being positioned prone in the MR-compatible animal holder. While scanning, respiration is monitored, and core body temperature is maintained at 37 °C using a warm air system (*see* **Note 4**).

### 3.1 Magnetic Resonance Imaging (MRI) Experiments for In Vivo Measurement of Whole-Body Fat Content and Distribution in Mouse

1. Horizontal 9.4 T scanner with a bore size of 31 cm.
2. A gradient system with a 12 cm inner diameter and a maximum gradient strength of 600 mT/m is used.
3. A volume coil with a 40 mm inner diameter and 110 mm RF-window is employed both for excitation and detection.
4. Axial slices of 1 mm thickness with 0.7 mm gap, covering the mouse from neck to tail, with a field-of-view (FOV) of $51.2 \times 51.2\ mm^2$ and a matrix size of $256 \times 96$ are acquired, using a spin-echo sequence preceded by a 2 ms gauss shaped pulse and followed by a 1.5 ms crusher gradient of 17 G/cm which efficiently suppresses water while leaving fat unperturbed.
5. Data acquisition is restricted to the exhalation period to avoid motion artifacts due to respiration.
6. In order to have fat-only images the spin-echo sequence is employed with a TE = 8 ms.
7. Preceding every excitation pulse, a 2 ms gauss pulse selective for water followed by a 1.5 ms crusher gradient of 17 G/cm is applied to saturate the water signal, while leaving the fat signal unperturbed.
8. Respiration gating is utilized, where 22 slices are excited for each exhalation period, resulting in an effective TR of two respiration periods of approximately 3 s and a total scan time around 5 min.
9. Signal from fat on each slide is then computed using Fiji software (http://fiji.sc) [11] to calculate total fat (TF), visceral fat (VAT), total subcutaneous fat (SAT), interscapular white fat (iWAT), and interscapular brown fat (iBAT) (Fig. 1) [12–15].
10. VAT is calculated as the difference between the TF signal and SAT signal in the abdominal region.

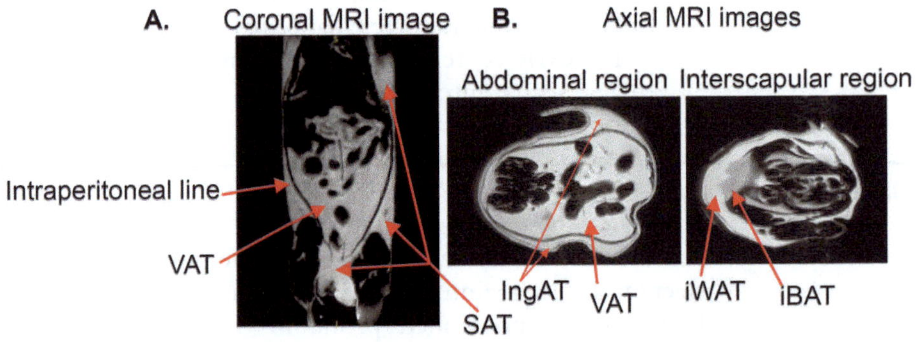


**Fig. 1** MR images of the adipose tissue obtained *in vivo* from a high-fat diet–mouse. (**a**) Whole-body coronal section and (**b**) abdominal and interscapular axial sections. *VAT* visceral adipose tissue, *SAT* subcutaneous adipose tissue, IngAT: Inguinal AT, *iWAT* interscapular white adipose tissue, *iBAT* interscapular brown adipose tissue

11. Inguinal fat (ingAT) counts for the SAT from the abdominal region (Fig. 1a).
12. iWAT is calculated as the difference between the TF signal and the iBAT signal in the interscapular region (Fig. 1b) [13].

### 3.2 Proton Magnetic Resonance Spectroscopy for In Vivo Evaluation of Hepatic Fat Content and Hepatic Lipid Composition in Mouse

As for the MRI experiment, animals are anesthetized using isoflurane (4% for induction of deep sleep and ~2% maintenance) in a 3:7 mixture of oxygen and air, before being positioned prone in the MR-compatible animal holder. While scanning, respiration is monitored, and core body temperature is maintained at 37 °C using a warm air system.

1. Proton magnetic resonance spectroscopy ($^1$H-MRS) is conducted using a horizontal 9.4 T scanner with a 31-cm bore diameter system.
2. $^1$H-MRS in liver is acquired using the point resolved spectroscopy [16, 17] with a volume coil of 40 mm inner diameter and an RF-window of 60 mm (*see* **Note 4**).
3. A flow-compensated gradient echo pulse sequence, with triggers to the first R-wave within the expiratory phase, is used for the obtention of the anatomical reference images, to localize the liver.
4. The field-of-view used is $60 \times 30$ mm$^2$ with a matrix dimension of $256 \times 128$.
5. 11 slices of 0.8 m thickness separated by 0.2 mm gaps with an effective $T_E$ of 2.8 ms are acquired.
6. To minimize the influence from respiration and heart motion, data acquisition is triggered from the R-wave of an ECG found in the exhalation phase.
7. Effective $T_R$ equals the respiration period (approximately 1 s), while the excitation and data acquisition of one k-space line from each of the 11 slices was accomplished in 47 ms.
8. A voxel size of $2 \times 2 \times 2$ mm$^3$ is then carefully placed in the left lobe of the liver to avoid contributions from blood vessels (Fig. 2).
9. A Shinnar Le Roux pulse of 1 ms is then used for excitation and Mao pulses of 1.1 ms are used for refocusing.
10. The carrier frequency is changed to 1200 Hz when acquiring the fat spectra (nex = 96) to avoid chemical shift displacement, and back to the water resonance when acquiring the water reference scan (nex = 8).
11. Three manually positioned saturation bands are used for outer volume suppression.

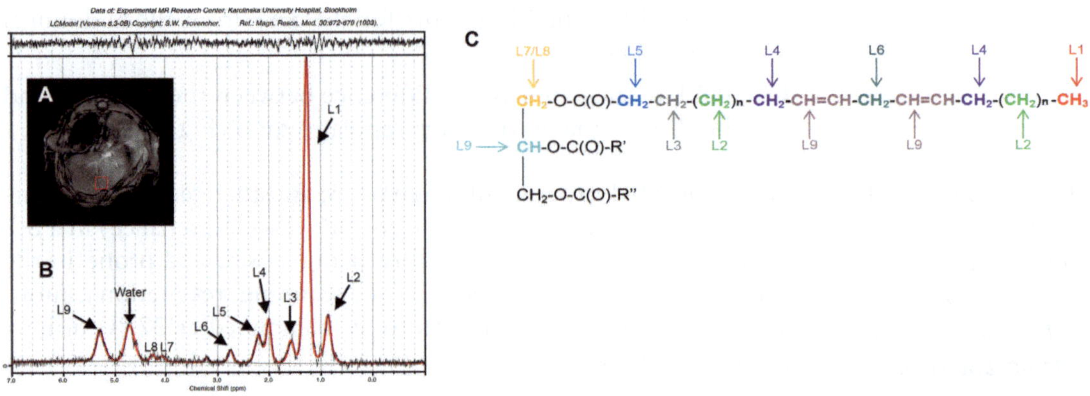


**Fig. 2** MRI image and $^1$H-MRS lipid spectrum obtained in vivo from a high-fat diet–mouse liver. (**a**) Axial image of the liver with voxel positioned; (**b**) representative spectrum from mouse liver; and (**c**) correspondence of the nine different 1H-MRS peaks to different 1H positions in triacylglycerol molecule. References are labelled according to their chemical shift values

12. All spectroscopy data are processed using LCModel analysis software (http://s-provencher.com/pub/LCModel/manual/manual.pdf) [18].
13. "Liver 8" is used as a base with all signals occurring in the spectral range of 0–7 ppm (water resonance at 4.7 ppm) simulated in LCModel.
14. All concentrations are derived from the area of the resonance peaks of the individual metabolites [13, 18].
15. Only the fitting results with an estimated standard deviation less than 20% are further analyzed.

Several parameters can then be derived from the intensity of the nine lipid signals normalized to the number of protons that contribute to the signal and the water signal intensity [16, 17]:

1. The lipid mass (LM).
2. The fraction of the lipid mass (fLM).
3. The unsaturated lipid component (UL).
4. The fraction of unsaturated lipids (fUL).
5. The fraction of polyunsaturated lipids (fPUL).
6. The fraction of monounsaturated lipids (fMUL).
7. The saturated lipid component (SL).
8. The fraction of saturated lipids (fSL).
9. The medium chain length.

*The lipid mass (LM) and the fraction of the lipid mass (fLM)*: Protons from water, mono-, di-, and triglycerides, phospholipids, and free fatty acids, as well as a small amount of cholesterol and cholesterol ester molecules mostly contribute to the $^1$H MR spectrum [20, 21]. Phospholipid contributions can be ignored, attributable to restricted mobility of these membrane lipids [21].

Therefore, the LM could be assessed from the weighted sum of all lipid signals with weighting factors 13 accounting for the mass of one =CH– group (L9), 14 for $-CH_2-$ group (L2, L3, L4, L5, L6, L7, and L8) and 15 for $-CH_3$ group (L1), if we assume one $-CH_3$ per fatty acid chain (Table 1). Accordingly, the lipid mass may be assessed as follows:

$$\begin{aligned}\text{Lipid mass} &= \frac{1}{2} \times (\text{L2} + \text{L3} + \text{L4} + \text{L5} + \text{L6} + \text{L7} + \text{L8}) \times 14 \\ &\quad + (\text{L9}) \times 13 + \frac{1}{3} \times (\text{L1}) \times 15 + \frac{1}{3} \times (\text{L1}) \times 15 \\ &\quad + \frac{1}{3} \times (\text{L1}) \times 12 + \frac{2}{3} \times (\text{L1}) \times 16\end{aligned} \tag{1}$$

with the ratios normalizing for the number of protons contributing to the signal. The last two terms in Eq. 1 include contributions for one carbonyl carbon (mass 12) and of two oxygen atoms (mass 16) per acid chain. The fLM is obtained as the mass ratio of the lipid mass divided by the sum of lipid plus water mass; hence, Eq. 2) is as follows:

$$\text{fLM} = \frac{\text{Lipid mass}}{\text{Lipid mass} + \frac{1}{2} \times I_{\text{water}} \times 18} \tag{2}$$

$I_{\text{water}}$ is the intensity of water signal, the factor $\frac{1}{2}$ stands for the two protons of the water signal, and 18 is the water molecular weight.

*The unsaturated lipid component* ($\text{UL}_{\text{total}}$) is defined as the average number of carbon double bonds per fatty acid chain (Table 1), that is,

$$\text{UL}_{\text{TOTAL}} = \frac{3}{2} \times \frac{\text{L9}}{\text{L1}} \tag{3}$$

The peak L9 contains contributions from both mono- and polyunsaturated lipids as well as the signal from glycerol [22], but the spectral resolution in vivo is not high enough to distinguish the various resonances. The relationship of $\text{L9} - \frac{1}{9} \times \text{L1}$ is applied to exclude the contribution from glycerol used in previous work by Giarola [23]. This is applicable with the assumption that there are only triglycerides in hepatic lipids, which is not correct. The factor 2 in the denominator accounts for the fact that $\text{UL}_{\text{total}}$ accounts for the number of double bonds and not the number of =CH– groups.

*The fraction of unsaturated lipids* (fUL) is the average percentage of fatty acid chains with double carbon bonds, including mono- and polyunsaturated fatty acids (Table 1), that is,

**Table 1**
**Nine lipid signals are identified in the murine liver $^1$H-MRS spectra**

| Peak | Chemical shift (ppm) | Chemical element–associated proton of the spectra |
|---|---|---|
| L1 | 0.90 | $\mathbf{CH_3}$–$(CH_2)_n$– |
| L2 | 1.30 | –$(\mathbf{CH_2})_n$– |
| L3 | 1.60 | –$CH_2$–O–CO–$CH_2$–$\mathbf{CH_2}$– |
| L4 | 2.03 | –$CH_2$–$\mathbf{CH_2}$–CH=CH– |
| L5 | 2.25 | –$CH_2$–O–CO–$\mathbf{CH_2}$–$CH_2$– |
| L6 | 2.76 | –CH=CH–$\mathbf{CH_2}$–CH=CH– |
| L7 | 4.09 | –$\mathbf{CH_2}$–O–C(O)–$CH_2$–$CH_2$– |
| L8 | 4.28 | –$\mathbf{CH_2}$–O–C(O)–$CH_2$–$CH_2$– |
| L9 | 5.32 | –**CH**=**CH**– and >**CH**–$CH_2$–O–C(O)–$CH_2$–$CH_2$– |

Peak assignments are as follows based on published data [19]

$$fUL = \frac{3}{4} \times \frac{L4}{L1} \quad (4)$$

with the factor $\frac{1}{4}$ accounts for the two methylene groups with two protons each, adjacent to both single and double bonds.

*The fraction of polyunsaturated lipids* (fPUL) is the percentage of fatty acid chains with multiple double bonds (Table 1), that is,

$$fPUL = \frac{3}{2} \times \frac{L6}{L1} \quad (5)$$

with the factor $\frac{1}{2}$ for the methylene group between two double bonds.

*The fraction of monounsaturated lipids* (fMUL), that is, the percentage of fatty acid chains containing a single carbon double bond (Table 1), is as follows:

$$fMUL = fUL - fPUL \quad (6)$$

*The saturated lipid component* (SL) is the average number of –$CH_2$– groups (L2) per fatty acid chain, which is obtained by normalization with L1 signal intensity as there is only one –$CH_3$ group per lipid chain. L1 instead of L3 or L5 is used for the quantification of number of chains (Table 1), since the L1 resonance is better separated from bordering peaks and therefore more reliable. Consequently, we obtain the following:

$$SL = \frac{3}{2} \times \frac{L2}{L1} \quad (7)$$

*The fraction of saturated lipid* f(SL) is the percentage of fatty acid chain without double bonds, that is,

$$fSL = 1 - fUL \tag{8}$$

The medium chain length (MCL) in then estimated as follows (Table 1):

$$MCL = \frac{\frac{1}{3} \times L1 + \frac{1}{2} \times (L2 + L3 + L4 + L5 + L6) + L9}{\frac{1}{3}L1} + 1 \tag{9}$$

where the term 1 stands for one carbonyl group per chain.

## 4 Notes

1. The *db/db* mouse model of leptin receptor deficiency is the most widely used mouse model of T2D. Experiments using *db/db* model are more widely conducted with the C57BlKS/J strain, as C57Bl/6 mice do not develop significant nephropathy [24].
2. There are a large variety of high-fat diets for rodent models. However, the composition and the quality of the diet may vary significantly between providers and may be a key factor in the discrepancies observed in the literature. In addition, according to the rules of your institution and your animal facility you may need to irradiate your diet once or twice before it can enter the animal facility; therefore, vitamin content must be adjusted accordingly.
3. MR scanners are designed for animal studies as for human studies. The main difference is the opening bore diameter that is higher for those used for human studies. All experimental setups described here are transferable to humans when adjusted for the sequences.
4. The sequence described in our study is based on a mouse model and with a horizontal 9.4 T scanner with a 31-cm bore diameter (Varian, Yarnton, UK) system. However, other types of scanners (Bruker, Siemens, etc.), gradient systems, and coils (surface or volume coils) may be used; the sequence must then be adjusted accordingly.

## Acknowledgments

I would like to thank the Department of Comparative Medicine/Karolinska Experimental Research and Imaging Centre at Karolinska University Hospital, Solna, Sweden for their support. In particular, I am very grateful to Sahar Nikkhou Aski and Peter Damberg for excellent assistance during the experiments and for their help in developing the sequences to perform optimal images and spectrum.

## References


1. Raatz SK, Conrad Z, Johnson LK, Picklo MJ, Jahns L (2017) Relationship of the reported intakes of fat and fatty acids to body weight in US adults. Nutrients 9(5). https://doi.org/10.3390/nu9050438
2. Lissner L, Heitmann BL (1995) Dietary fat and obesity: evidence from epidemiology. Eur J Clin Nutr 49(2):79–90
3. Wang CY, Liao JK (2012) A mouse model of diet-induced obesity and insulin resistance. Methods Mol Biol 821:421–433. https://doi.org/10.1007/978-1-61779-430-8_27
4. Schwab U, Lauritzen L, Tholstrup T, Haldorssoni T, Riserus U, Uusitupa M, Becker W (2014) Effect of the amount and type of dietary fat on cardiometabolic risk factors and risk of developing type 2 diabetes, cardiovascular diseases, and cancer: a systematic review. Food Nutr Res 58. https://doi.org/10.3402/fnr.v58.25145
5. Liu X, Li Y, Tobias DK, Wang DD, Manson JE, Willett WC, Hu FB (2018) Changes in types of dietary fats influence long-term weight change in US women and men. J Nutr 148 (11):1821–1829. https://doi.org/10.1093/jn/nxy183
6. Zacek P, Bukowski M, Mehus A, Johnson L, Zeng H, Raatz S, Idso JP, Picklo M (2019) Dietary saturated fatty acid type impacts obesity-induced metabolic dysfunction and plasma lipidomic signatures in mice. J Nutr Biochem 64:32–44. https://doi.org/10.1016/j.jnutbio.2018.10.005
7. Krishnan S, Cooper JA (2014) Effect of dietary fatty acid composition on substrate utilization and body weight maintenance in humans. Eur J Nutr 53(3):691–710. https://doi.org/10.1007/s00394-013-0638-z
8. Estrany ME, Proenza AM, Llado I, Gianotti M (2011) Isocaloric intake of a high-fat diet modifies adiposity and lipid handling in a sex dependent manner in rats. Lipids Health Dis 10:52. https://doi.org/10.1186/1476-511X-10-52
9. Coleman DL, Hummel KP (1967) Studies with the mutation, diabetes, in the mouse. Diabetologia 3(2):238–248
10. Coleman DL, Hummel KP (1973) The influence of genetic background on the expression of the obese (Ob) gene in the mouse. Diabetologia 9(4):287–293
11. Image processing and analysis with ImageJ and MRI Cell Image Analyzer (2010) photologweb.com
12. Archer A, Stolarczyk E, Doria ML, Helguero L, Domingues R, Howard JK, Mode A, Korach-Andre M, Gustafsson JA (2013) LXR activation by GW3965 alters fat tissue distribution and adipose tissue inflammation in ob/ob female mice. J Lipid Res 54 (5):1300–1311. https://doi.org/10.1194/jlr.M033977
13. Gonzalez-Granillo M, Savva C, Li X, Fitch M, Pedrelli M, Hellerstein M, Parini P, Korach-Andre M, Gustafsson JA (2019) ERbeta activation in obesity improves whole body metabolism via adipose tissue function and enhanced mitochondria biogenesis. Mol Cell Endocrinol 479:147–158. https://doi.org/10.1016/j.mce.2018.10.007
14. Korach-Andre M, Gao J, Gounarides JS, Deacon R, Islam A, Laurent D (2005) Relationship between visceral adiposity and intramyocellular lipid content in two rat models of insulin resistance. Am J Physiol Endocrinol Metab 288(1):E106–E116. https://doi.org/10.1152/ajpendo.00089.2004
15. Korach-Andre M, Parini P, Larsson L, Arner A, Steffensen KR, Gustafsson JA (2010) Separate and overlapping metabolic functions of LXRalpha and LXRbeta in C57Bl/6 female mice. Am J Physiol Endocrinol Metab 298(2): E167–E178. https://doi.org/10.1152/ajpendo.00184.2009
16. Ye Q, Danzer CF, Fuchs A, Vats D, Wolfrum C, Rudin M (2013) Longitudinal evaluation of hepatic lipid deposition and composition in ob/ob and ob/+ control mice. NMR Biomed 26(9):1079–1088. https://doi.org/10.1002/nbm.2921
17. Ye Q, Danzer CF, Fuchs A, Wolfrum C, Rudin M (2012) Hepatic lipid composition differs between ob/ob and ob/+ control mice as determined by using in vivo localized proton magnetic resonance spectroscopy. MAGMA 25 (5):381–389. https://doi.org/10.1007/s10334-012-0310-2
18. Hajek M, Burian M, Dezortova M (2000) Application of LCModel for quality control and quantitative in vivo 1H MR spectroscopy by short echo time STEAM sequence. MAGMA 10(1):6–17
19. Provencher SW (1993) Estimation of metabolite concentrations from localized in vivo proton NMR spectra. Magn Reson Med 30 (6):672–679
20. Bollard ME, Garrod S, Holmes E, Lindon JC, Humpfer E, Spraul M, Nicholson JK (2000)

High-resolution (1)H and (1)H-(13)C magic angle spinning NMR spectroscopy of rat liver. Magn Reson Med 44(2):201–207

21. Ruberg FL, Viereck J, Phinikaridou A, Qiao Y, Loscalzo J, Hamilton JA (2006) Identification of cholesteryl esters in human carotid atherosclerosis by ex vivo image-guided proton MRS. J Lipid Res 47(2):310–317. https://doi.org/10.1194/jlr.M500431-JLR200
22. Ren J, Dimitrov I, Sherry AD, Malloy CR (2008) Composition of adipose tissue and marrow fat in humans by 1H NMR at 7 Tesla. J Lipid Res 49(9):2055–2062. https://doi.org/10.1194/jlr.D800010-JLR200
23. Giarola M, Rossi B, Mosconi E, Fontanella M, Marzola P, Scambi I, Sbarbati A, Mariotto G (2011) Fast and minimally invasive determination of the unsaturation index of white fat depots by micro-Raman spectroscopy. Lipids 46(7):659–667. https://doi.org/10.1007/s11745-011-3567-8
24. Kong LL, Wu H, Cui WP, Zhou WH, Luo P, Sun J, Yuan H, Miao LN (2013) Advances in murine models of diabetic nephropathy. J Diabetes Res 2013:797548. https://doi.org/10.1155/2013/797548

# Chapter 13

# An Obesogenic Dietary Mouse Model of Nonalcoholic Steatohepatitis

**Bader Zarrouki and Jeremie Boucher**

## Abstract

Nonalcoholic steatohepatitis (NASH) is a severe form of nonalcoholic fatty liver disease (NAFLD), characterized by steatosis (fat within the liver), inflammation, and fibrosis, which may progress to cirrhosis and hepatocellular carcinoma. Despite the high prevalence, there are currently no approved NASH drug treatments, which urges a faster development of new therapies to address this high unmet medical need. Drug development is facilitated by having reliable and translatable preclinical NASH models. Obesogenic dietary models recapitulate better the natural progression of NASH, with overnutrition and sedentary lifestyle being the main causes. Here we describe the use of a modified version of a diet-induced NASH model, known as the Amylin NASH diet model (AMLN-diet), particularly in the leptin-deficient $Lep^{ob}$/ $Lep^{ob}$ (*ob/ob*) mice.

**Key words** NAFLD-NASH, Mouse model, Steatosis, Inflammation, Fibrosis, Ballooning-NAS

## 1 Introduction

Nonalcoholic steatohepatitis (NASH) is a severe form of nonalcoholic fatty liver disease (NAFLD), characterized by steatosis (fat within the liver), inflammation, and fibrosis, which may progress to cirrhosis and hepatocellular carcinoma [1]. NASH is the most common liver disorder and is becoming the leading cause of liver transplant [2, 3]. Despite the high prevalence, there are currently no approved NASH drug treatments [4]. Therefore, it is critical to develop new therapies to address this high unmet medical need. Drug development is facilitated by having reliable and translatable preclinical models. Several NASH mouse models have been developed recently [5]. Chemically induced models, such as carbon tetrachloride treatment, and murine models with diets deficient in methionine and choline cause liver damage and fibrosis and are most useful to study hepatic fibrosis development and regression [5, 6]. However, animals exposed to these interventions are not representative of the disease etiology, since they are usually not

Myriam Aouadi and Valerio Azzimato (eds.), *Kupffer Cells: Methods and Protocols*, Methods in Molecular Biology, vol. 2164,
https://doi.org/10.1007/978-1-0716-0704-6_13, © Springer Science+Business Media, LLC, part of Springer Nature 2020

obese, nor do they display insulin resistance and in most cases are in negative energy balance [5, 6]. Obesogenic dietary models on the other hand, recapitulate better the natural progression of NASH, with overnutrition and sedentary lifestyle being the main causes. A wide range of mouse genetic backgrounds and diets have been described, with varying kinetics and degree of NASH development [5–8]. Here we describe the use of a modified version of a diet-Induced NASH model, known as the Amylin NASH model diet (AMLN-diet), particularly in the leptin deficient $Lep^{ob}/Lep^{ob}$ (*ob/ob*) mice. Indeed, when fed a regular chow diet, these obese, hyperphagic, and extremely insulin resistant mice, develop dramatic liver steatosis but do not spontaneously progress to steatohepatitis or display detectable fibrosis [7, 8]. Interestingly, when fed the AMLN-diet, a high fat diet with 40% of calories from fat, particularly rich in trans-fat (Primex hydrogenated fat; 22% by weight), fructose (20% by weight) and cholesterol (2% by weight), *ob/ob* mice develop key features of NASH with a gradual onset of inflammation and reasonable fibrosis after 12 weeks of feeding [7, 8]. More recent reports [9, 10] have shown that 1- this diet is also able to drive disease onset in regular C57B mice, albeit the diet intervention time needed to elicit NASH hallmarks is significantly longer compared to *ob/ob* mice on the same diet and 2-clinical drug candidates for the treatment of NASH attenuates disease progression in these models. The recent FDA ban on trans-fats in foods hampered the availability and use of the AMLN diet. However, a replacement diet, with a composition and caloric content similar to the AMLN diet, where primex trans-fat was replaced by palm oil, has been shown to be capable of promoting a similar level of disease as the AMLN diet [11]. Herein we provide the detailed protocol of this diet-induced NASH model and recommendations when performing drug intervention studies in this model.

## 2 Materials

### *2.1 Animals*

| Species | Mouse |
|---|---|
| Strain | B6.Cg-Lepob/J (ob/ob), C57BL/6 J (lean) |
| Sex | Males |
| Age at arrival to facilities | 6–7 weeks |
| Housing conditions | *ob/ob* mice are housed 4 mice per cage<br>Lean animals are housed individually |

(continued)

| Species | Mouse |
|---|---|
| | Cages with Aspen wood chip bedding (Tapvei Oy, Kortteinen, Finland). A little house, nesting material, and chew sticks are provided<br>12 h light–dark cycle, room temperature 22 °C, and relative humidity 40–60%<br>Tap water and diet are given ad libitum |

### 2.2 Diet

1. Regular chow diet (R3; Lactamin AB, Kimstad, Sweden).
2. NASH diet (CAT#: D09100310, Research diets, New Brunswick, NJ, USA): this is a high-fat diet with 40% of calories from fat (mostly palm oil), rich in fructose (22%) and 2% cholesterol (2%).

## 3 Methods

### 3.1 Diet and Treatment Intervention

1. After 1 week acclimatization upon arrival, randomize mice based on body weight and start feeding ob/ob mice with either NASH or chow for a minimum of 16 weeks [11]. Lean C57BL/6 J mice fed show control diet can be used to get reference values of under healthy status (*see* **Note 1**).
2. Monitor body weight weekly.
3. Circulating glucose, insulin, alanine aminotransferase (ALT) and aspartate aminotransferase (AST), monthly through tail vein sampling in restrained, conscious animals.
4. Before starting drug treatment, randomize animals into different treatment groups based on body weight, ALT and AST (*see* **Note 2**).
5. Drug treatment (*see* **Note 3**):
   (a) NASH prevention mode: initiate treatment simultaneously to initiation of diet intervention up to 16 weeks on diet.
   (b) NASH reversal mode: initiate treatment when mice have been on diet for at least at 16 weeks [11].

### 3.2 Termination

1. We recommend starving the mice 3–4 h before termination.
2. Anesthetize mice with a mixture of isoflurane (2–3%) in 100% oxygen.
3. Quickly withdraw blood through retro-orbital bleeding into EDTA-coated tubes (*see* **Note 4**).
4. Subsequently mice should be terminated by exsanguination through cardiac bleed under anesthesia or cervical dislocation. All macroscopic abnormalities should be recorded.

5. Perform laparotomy to harvest liver for histological analysis. A detailed procedure is described in the next section.

### 3.3 Tissue Harvesting Preparation and Histopathology

1. Harvest a piece of left lateral lobe (~100 mg) encompassing the edge and deeper portion of the liver (*see* **Note 5**).
2. Put the liver piece in 4% neutral buffered formaldehyde for 48 h and subsequently embed in paraffin.
3. Four μm thick formaldehyde-fixed, paraffin-embedded mouse liver adjacent sections should be stained as below to assess disease end points (*see* **Note 6**):
4. Hematoxylin–eosin (H&E) for steatosis, ballooning, and inflammation staging.
5. Pico-Sirius Red staining for fibrosis staging.
6. Adjacent sections can be immunohistochemically stained for markers of inflammation (Mac-2) and fibrosis (Col1A1) (*see* **Note 7**).

### 3.4 Biochemical Assays

Plasma clinical chemistry and biochemical assays should be analyzed adequately. We perform these analyses as described below and in [12]:

1. In nonanesthetized animals, blood glucose and insulin levels are measured in tail vein blood using a portable glucometer (Accu-Chek mobile®) and the ultrasensitive mouse insulin ELISA Kit (#90080, Crystal Chemical, Downers Grove, IL, USA).
2. Upon termination: Plasma insulin levels are measured using a mouse/rat Insulin kit from Meso Scale Discovery (K152BZC-1, Rockville Maryland USA). Plasma triglyceride, glucose, ALT, AST, cholesterol and haptoglobin levels are analyzed using an ABX Pentra 400 instrument (Horiba Medical, Irvine, California, USA).
3. Liver hydroxyproline is assessed in a frozen liver piece (~100 mg) using the Hydroxyproline Assay kit (#QZBHYPRO1, QuickZyme Biosciences, Leiden, The Netherlands), according to the manufacturer's protocol.
4. Liver triglyceride content is assessed in a frozen liver piece (~50 mg) using an ABX Pentra 400 instrument (Horiba Medical, Irvine, California, USA) and detailed processing is described here [12].

## 4 Notes

1. Diet should be changed weekly. We advise to run a time course characterization study when performing this procedure for the

first time to generate relevant tissue and plasma samples to evaluate the model and investigate the regulation of target/pathway of interest. In addition, this will provide data to perform adequate power calculation based on relevant primary endpoints. NASH can also develop in C57BL/6 J mice fed NASH diet, but this requires a longer duration of diet treatment to detect onset of disease [7–9].

2. Sample size should be determined through statistical assessment and power calculation. In our experience a minimum of 12–15 animals/group should be used when investigating treatment effect on fibrosis. However, as for any diet-induced model, it is anticipated that mice will have different degree of disease progression and this variation might be a pitfall in assessing treatment effect [8]. It has been suggested that performing prebiopsy (described in [8–11, 13]) where disease progression is assessed by histopathology and randomization done using NAS score and fibrosis staging, in a way similar to a clinical setup, represents a substantial improvement of the use of this model. However, this means that at least one third of animals will not be included in the study based on this procedure [8, 9]. A filmed version of this procedure has been published recently [13].
3. Regarding the duration of drug treatment. It has been documented that treatment effect can be seen as early as 8 weeks upon intervention in this model [10]; however based on our experience we advise a treatment duration between 12 and 16 weeks if the primary end point is to assess effect on fibrosis staging. The indicated time of treatment start in this chapter is based on the use of the *ob/ob* mice where disease onset occurs earlier than regular C57BL/6 mice, which require a longer time on diet to develop disease features [8–11, 13].
4. We do not recommend performing heart puncture since damaged cardiac muscle may release AST which would be a confounding factor.
5. Other liver lobes, such as the right medial lobe or left medial lobe could be considered. However, it is important to be consistent with location, size, and pattern of the harvested liver tissue within and between studies.
6. The gold standard method for assessing NASH disease status and progression should be histological assessment and scoring. This should be performed by a pathologist blinded to the study and familiar with NASH disease assessment. NAFLD activity score (NAS) (steatosis/inflammation/ballooning degeneration) and fibrosis stage should be performed according to the clinical considerations outlined by Kleiner et al. [14]. Based on our assessment, we find no or little ballooning in this model,

therefore we do not consider this model as a good model to assess ballooning. Image quantification can also be performed through scanning of the whole stained sections at ×20 magnification and digital analysis of whole sections can be performed using several software (Image J, Visiomorph, Biopix). In our case, digital image analysis is done by Visiomorph software (Visiopharm, Copenhagen, Denmark). Subsequently, hepatic steatosis, inflammation and fibrosis can be presented as % of total area.

7. These antibodies can be used to stain for markers of macrophages (Mac-2, CL8942AP Cedarlaine, Canada) and collagen 1 (Col1A1, LS-C343921, BioSite, USA).

## Disclosure

The authors are AstraZeneca employees and declare no competing interests related to this work.

### References

1. Diehl AM, Day C (2017) Cause, pathogenesis, and treatment of nonalcoholic steatohepatitis. N Engl J Med 377(21):2063–2072. https://doi.org/10.1056/NEJMra1503519
2. Bellentani S (2017) The epidemiology of non-alcoholic fatty liver disease. Liver Int 37 (Suppl 1):81–84. https://doi.org/10.1111/liv.13299
3. Estes C, Anstee QM, Arias-Loste MT, Bantel H, Bellentani S, Caballeria J, Colombo M, Craxi A, Crespo J, Day CP, Eguchi Y, Geier A, Kondili LA, Kroy DC, Lazarus JV, Loomba R, Manns MP, Marchesini G, Nakajima A, Negro F, Petta S, Ratziu V, Romero-Gomez M, Sanyal A, Schattenberg JM, Tacke F, Tanaka J, Trautwein C, Wei L, Zeuzem S, Razavi H (2018) Modeling NAFLD disease burden in China, France, Germany, Italy, Japan, Spain, United Kingdom, and United States for the period 2016-2030. J Hepatol 69(4):896–904. https://doi.org/10.1016/j.jhep.2018.05.036
4. Issa D, Patel V, Sanyal AJ (2018) Future therapy for non-alcoholic fatty liver disease. Liver Int 38(Suppl 1):56–63. https://doi.org/10.1111/liv.13676
5. Hansen HH, Feigh M, Veidal SS, Rigbolt KT, Vrang N, Fosgerau K (2017) Mouse models of nonalcoholic steatohepatitis in preclinical drug development. Drug Discov Today 22 (11):1707–1718. https://doi.org/10.1016/j.drudis.2017.06.007
6. Larter CZ, Yeh MM (2008) Animal models of NASH: getting both pathology and metabolic context right. J Gastroenterol Hepatol 23 (11):1635–1648. https://doi.org/10.1111/j.1440-1746.2008.05543.x
7. Trevaskis JL, Griffin PS, Wittmer C, Neuschwander-Tetri BA, Brunt EM, Dolman CS, Erickson MR, Napora J, Parkes DG, Roth JD (2012) Glucagon-like peptide-1 receptor agonism improves metabolic, biochemical, and histopathological indices of nonalcoholic steatohepatitis in mice. Am J Physiol Gastrointest Liver Physiol 302(8):G762–G772. https://doi.org/10.1152/ajpgi.00476.2011
8. Clapper JR, Hendricks MD, Gu G, Wittmer C, Dolman CS, Herich J, Athanacio J, Villescaz C, Ghosh SS, Heilig JS, Lowe C, Roth JD (2013) Diet-induced mouse model of fatty liver disease and nonalcoholic steatohepatitis reflecting clinical disease progression and methods of assessment. Am J Physiol Gastrointest Liver Physiol 305(7):G483–G495. https://doi.org/10.1152/ajpgi.00079.2013
9. Kristiansen MN, Veidal SS, Rigbolt KT, Tolbol KS, Roth JD, Jelsing J, Vrang N, Feigh M (2016) Obese diet-induced mouse models of nonalcoholic steatohepatitis-tracking disease by liver biopsy. World J Hepatol 8 (16):673–684. https://doi.org/10.4254/wjh.v8.i16.673
10. Tolbol KS, Kristiansen MN, Hansen HH, Veidal SS, Rigbolt KT, Gillum MP, Jelsing J,

Vrang N, Feigh M (2018) Metabolic and hepatic effects of liraglutide, obeticholic acid and elafibranor in diet-induced obese mouse models of biopsy-confirmed nonalcoholic steatohepatitis. World J Gastroenterol 24 (2):179–194. https://doi.org/10.3748/wjg.v24.i2.179

11. Boland ML, Oro D, Tolbol KS, Thrane ST, Nielsen JC, Cohen TS, Tabor DE, Fernandes F, Tovchigrechko A, Veidal SS, Warrener P, Sellman BR, Jelsing J, Feigh M, Vrang N, Trevaskis JL, Hansen HH (2019) Towards a standard diet-induced and biopsy-confirmed mouse model of non-alcoholic steatohepatitis: impact of dietary fat source. World J Gastroenterol 25(33):4904–4920. https://doi.org/10.3748/wjg.v25.i33.4904

12. Linden D, Ahnmark A, Pingitore P, Ciociola E, Ahlstedt I, Andreasson AC, Sasidharan K, Madeyski-Bengtson K, Zurek M, Mancina RM, Lindblom A, Bjursell M, Bottcher G, Stahlman M, Bohlooly YM, Haynes WG, Carlsson B, Graham M, Lee R, Murray S, Valenti L, Bhanot S, Akerblad P, Romeo S (2019) Pnpla3 silencing with antisense oligonucleotides ameliorates nonalcoholic steatohepatitis and fibrosis in Pnpla3 I148M knock-in mice. Mol Metab 22:49–61. https://doi.org/10.1016/j.molmet.2019.01.013

13. Oldham S, Rivera C, Boland ML, Trevaskis JL (2019) Incorporation of a survivable liver biopsy procedure in mice to assess non-alcoholic steatohepatitis (NASH) resolution. J Vis Exp (146). https://doi.org/10.3791/59130

14. Kleiner DE, Brunt EM, Van Natta M, Behling C, Contos MJ, Cummings OW, Ferrell LD, Liu YC, Torbenson MS, Unalp-Arida A, Yeh M, McCullough AJ, Sanyal AJ, Nonalcoholic Steatohepatitis Clinical Research Network (2005) Design and validation of a histological scoring system for nonalcoholic fatty liver disease. Hepatology 41 (6):1313–1321. https://doi.org/10.1002/hep.20701

# Chapter 14

# A Transposon-Based Mouse Model of Hepatocellular Carcinoma via Hydrodynamic Tail Vein Injection

Sijia Yu and Santiago Vernia

## Abstract

Transgenic mouse are reliable, convenient models for studying human hepatocellular carcinoma (HCC). The development of a synthetically engineered Sleeping Beauty (SB) transposon system further enables the viral-free, efficient delivery of desired oncogenes to mouse tissues. Here, we describe an SB transposon-based approach to induce HCC in mice by expressing a hyperactive form of N-RAS, N-RAS$^{G12V}$, while silencing the endogenous *Trp53* gene via hydrodynamic tail vein injection, a method to rapidly deliver naked plasmids to mouse liver.

**Key words** Hepatocellular carcinoma, Mouse tumor model, N-RAS, p53, Sleeping Beauty, Hydrodynamic tail vein injection

## 1 Introduction

Hepatocellular carcinoma (HCC) is the most common form of primary liver cancer and has become the fourth leading cause of cancer death worldwide in 2018 [1]. Hepatitis B virus (HBV) infection, along with aflatoxin B1 exposure, is the main risk factor to HCC development in sub-Saharan Africa and East Asia which contributed to 80% of the global HCC cases [2].

In studies of human HCC, mouse models are widely used due to their similar genetics and physiology to human, as well as the short life span and high breeding capacity [3]. Nowadays, more and more manipulating techniques have also been developed, which makes it more convenient to perform experiments and to generate mouse tumor models with various genetic backgrounds. Hydrodynamic tail vein injection (HTVI) is one of the most commonly used techniques for genetic modification, mainly for mouse liver. It works by rapidly injecting large volume of solution, allowing the permeabilization of capillary wall by hydrodynamic pressure, resulting in the delivery of experimental materials within the solution into the interior of the cells [4]. Hepatocytes in pericentral regions

Myriam Aouadi and Valerio Azzimato (eds.), *Kupffer Cells: Methods and Protocols*, Methods in Molecular Biology, vol. 2164, https://doi.org/10.1007/978-1-0716-0704-6_14, © Springer Science+Business Media, LLC, part of Springer Nature 2020

are most likely to be affected, and a heterogeneous population of cells carrying the exogenous genes delivered is usually observed [4]. This is advantageous compared to the traditional transgenic mouse model where the mutations are ubiquitously expressed in all cells of the body or liver, as it can better mimic tumor initiation in reality from single cells acquired abnormalities [5].

In addition, a synthetically engineered SB transposon system was generated in 1997 to integrate and stably express oncogenes in mouse germline and somatic tissues including liver, and is particularly advantageous owing to its ability of targeting nondividing cells [6–8]. The reconstructed SB transposon activates a silenced gene encoding Sleeping Beauty transposase (SB10) that is equivalent to the ancient *Tc1/mariner* superfamily transposase in salmonid. SB10 binds and cuts at the inverted/direct repeats (IR/DRs) flanking the target gene in salmonid transposons, followed by integrating the gene into genome where there is a TA site to provoke stable expression [8]. More recently, another version of the Sleeping Beauty transposase, SB13, was generated from SB10. With two point mutations (T83 and K33A), the transpositional activity of SB13 has been greatly enhanced compared to that of SB10 [9].

Over the years, multiple transgenic, chemical-induced, or diet-induced rodent models have been created for the study of HCC [10]. Among these, an SB transposon-based mouse model that expresses a constitutively active form of N-RAS, N-RAS$^{G12V}$, along with the silence of *Trp53* gene, has been proven to be a fit model for human HBV-induced HCC [11].

N-RAS is a member of the RAS small GTPase family, activating genes involved in cell growth and survival. Activating mutations in *N-Ras* gene, including G12V, are found in about 30% of the human liver carcinoma [12]. Expression of N-RAS$^{G12V}$ in mouse liver is associated with hyperactivation of the proproliferative signaling pathways, such as MAPK and/or PI3K/Akt pathways, as well as oncogene induced senescence in the first few weeks of expression [11, 13]. Although N-RAS$^{G12V}$ alone is sufficient enough to induce hyperplasia in mouse liver, additional approaches, such as simultaneous silence of the tumor-suppressor gene *Trp53*, are often used to accelerate the procedure [11].

Mouse *Trp53* gene (human homolog *TP53*) encodes a 53-kDa nuclear phosphoprotein which is considered as one of the most crucial tumor suppressors in organisms due to its ability to maintain genomic stability. Upon DNA damage, p53 initiates DNA repair pathways, pauses cell cycle and induces apoptosis if the damage persists, by inducing the expression of several genes, including *CDKN1A* (p21), the proapoptotic Bcl-2 family members and miR-34a [14]. Genomic analysis also reveals that *TP53* mutations are the second most frequent mutations in human HCC and found in about 45% of HCC patients with HBV infection [15, 16].

Here we describe the procedures to rapidly generate HCC in mice using the Sleeping Beauty (SB) transposon system to integrate *N-Ras*$^{G12V}$, a gene that encodes the oncogenic form of N-RAS, into mouse genome, while silencing the endogenous expression of tumor-suppressor gene *Trp53*, by hydrodynamic tail vein injection (HTVI). The identity of the nodules formed in mice liver is confirmed by a combination of typical morphological changes and expression of biomarkers for HCC using both histopathological and immunohistochemical studies.

## 2 Materials

### 2.1 For Endotoxin-Free plasmid Preparation

1. Plasmid (transposase plasmid and transgene plasmid).
   (a) Transposase plasmid: pPGK-SB13 (Addgene plasmid # 20207; http://n2t.net/addgene:20207; RRID: Addgene_20207) [17] (Fig. 1).
   (b) Transgene plasmid 1: pCaggs-*NRas*$^{G12V}$-*Gfp-shTrp53* that contains coding regions for N-RAS$^{G12V}$, GFP, and a miR-30-based short hairpin RNA (shRNA) that targets *Trp53*, adopted from the layouts in previously published papers [11, 13, 18] (Fig. 1).
   (c) Transgene plasmid 2 (negative control): pCaggs-*NRas*$^{G12VD38A}$-*Gfp-ShNC* that contains coding region for N-RAS$^{G12V/D38A}$, where the inactivation mutation D38A makes N-RAS unable to activate its downstream signaling cascade [13]. A noncoding shRNA is also included as the negative control [18] (Fig. 1).
2. Competent bacteria (e.g., *E. coli* TOP10 strain).
3. Sterile distilled water (d$H_2O$).
4. SOC media (350 μl per plasmid).
5. Sterile LB broth with 100 μg/ml ampicillin.
6. Sterile LB agar plates (10-cm diameter, containing 100 μg/ml ampicillin).
7. Qiagen EndoFree Maxi kit.
8. Safe-Lock microcentrifuge tubes.
9. Heat block.
10. Shake incubator.
11. Stationary incubator.
12. Ice bucket filled with ice.
13. Sterile spreading device.
14. Bacterial culture tube.
15. Conical flasks.

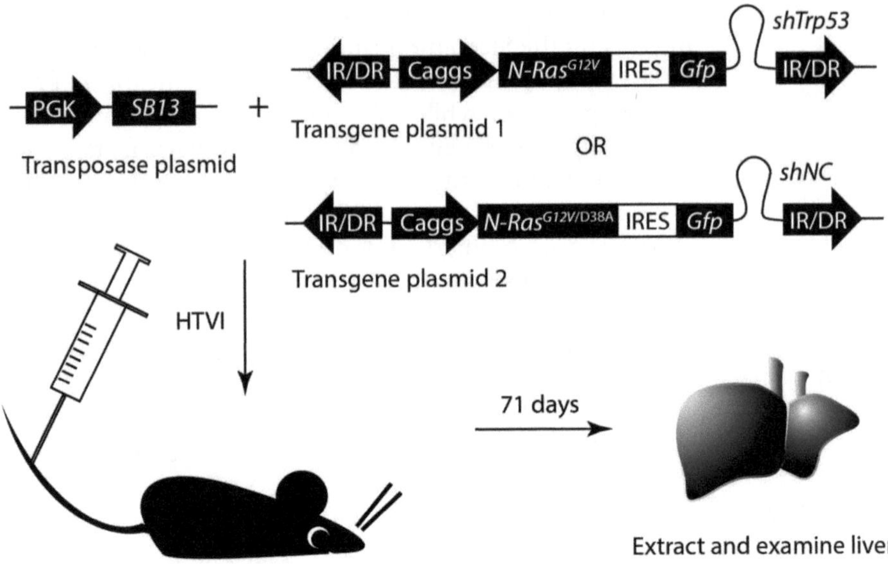


**Fig. 1** Schematic diagram for experiments containing design for plasmids. *PGK* Mouse phosphoglycerate kinase 1 promoter, *SB13* Sleeping Beauty transposase 13, *IR/DR* inverted/direct repeats, *Caggs* CAGGS promoter, *IRES* internal ribosome entry site, *GFP* green fluorescence protein, *shTrp53 and shNC* shRNAs targeting *Trp53* or noncoding sequence, *HTVI* hydrodynamic tail vein injection

16. Centrifuge tubes (500 ml).
17. Large-capacity, high-speed centrifuge.
18. Endotoxin-free microcentrifuge tubes and pipette tips.
19. UV-Vis spectrophotometers.

### 2.2 For Injection

1. C57BL/6 J mice (4–6 weeks old, 8 females and 8 males per condition).
2. Sterile saline (0.9% NaCl solution).
3. Plasmids from maxi-prep.
4. EMLA cream for local anesthesia.
5. Scale.
6. Sterile syringes.
7. Sterile needles.
8. Restrainer to hold mice.
9. Heat box.
10. Gauze swabs.
11. Towel roll or tissue.

### 2.3 For Evaluating Tumor Formation

1. Ethanol (100%, 95%, 80%, 70% and 50%, v/v). Diluted from ethanol absolute in $dH_2O$.

2. Tap water.
3. 10% Formalin solution, neutral buffered (NBF).
4. Phosphate buffered saline (PBS, 1×).
5. Xylenes.
6. Paraffin wax.
7. Mayer's hematoxylin solution (50 mg/ml aluminum ammonium sulfate, 1.2 mg/ml hematoxylin crystals, 0.2 mg/ml sodium iodate, 1.0 mg/ml citric acid, and 50 mg/ml chloral hydrate in double distilled water or purchase from Sigma, USA).
8. Eosin Y solution, alcoholic (Dissolve 2.0 g eosin Y in 40 ml double distilled water and add 160 ml 95% v/v ethanol to prepare the stock solution. When using, dilute 1:3 with 80% ethanol containing 0.5% glacial acetic acid. Alternatively, purchase from Sigma, USA).
9. Scotts tap water substitute concentrate 10× (2.0 mg/ml sodium bicarbonate, 20 mg/ml anhydrous magnesium sulfate in tap water or purchase from Sigma, USA).
10. Permount™ Mounting Medium (Fisher Scientific).
11. Citric acid buffer (10 mM citric acid monohydrate in $dH_2O$, pH = 6.0).
12. Hydrogen peroxide (30% w/w in $H_2O$).
13. StartingBlock™ T20 (PBS) Blocking Buffer (Thermo Scientific).
14. 1× BSA/PBT: Prepare 10% w/v BSA solution by dissolving 10 g BSA powder in $dH_2O$. Then, mix 5 ml PBS 10× with 0.5 ml BSA solution (10% w/v) and 1 ml Triton X-100 in 43.5 ml $dH_2O$.
15. Primary antibodies.
    (a) Anti-Ki67 (Abcam).
    (b) Anti-glypican 3 (Fisher Scientific).
    (c) Anti-p53 (Cell Signaling).
    (d) Anti-NRAS 1:500 (Santa Cruz Biotechnology).
16. VECTASTAIN® Elite® ABC HRP Kit (Peroxidase, Standard, Vector Laboratories).
17. DAB Peroxidase (HRP) Substrate Kit (with Nickel, Vector Laboratories).
18. Thermo Scientific™ Shandon™ Gill™ Hematoxylin (Fisher Scientific).
19. Destaining solution for hematoxylin staining (0.5% v/v hydrochloric acid in 70% v/v ethanol).

20. Molds for paraffin blocks.
21. Surgical tools (scissors, tweezers, scalpel, and working pads).
22. Scale.
23. Stereomicroscope for solid tumors and optical microscope for tissues slides.
24. Tissue cassettes.
25. Cold room at 4 °C.
26. Plastic slide rack.
27. Hybridization box.
28. Hydrophobic pen.
29. Tissue embedding station.
30. Microtome.
31. Water bath.
32. Microscopy slides.
33. Oven (heat up to 40 °C).

## 3 Methods

### 3.1 Plasmid Preparation

#### 3.1.1 Heat Shock Transformation

1. Transfer 100 ng of each plasmid to 100 μl competent bacteria in a microcentrifuge tube under sterile conditions. A negative control is made by transfer the same volume of sterile water to competent bacteria and an empty vector containing ampicillin resistance gene is used as a positive control. Mix by gently tapping the tube with fingers.
2. Incubate the mixture on ice for 20 min.
3. Heat-shock each tube by warming it up in heat block at 42 °C for 1 min.
4. Transfer the tube back on ice for 5 min.
5. Allow the bacteria to recover by adding 350 μl SOC media to each tube.
6. Incubate the tubes in a shake incubator at 37 °C for an hour.
7. Spread some or all the bacteria–plasmid mixture onto a 10-cm LB agar plate containing 100 μg/ml ampicillin (*see* **Note 1**).
8. Incubate the plates at 37 °C in a stationary incubator overnight.
9. Examine the colony formation in the following day. There should be no colony on the negative control plate as an indication of sterile handling and some colonies on the positive control plate as an indication of successful transformation.

#### 3.1.2 Maxi-Prep Culture Setup

1. Pick a single colony up from the transgene and transposase plasmid plate using a pipette tip under sterile conditions.
2. Transfer each tip to a bacterial culture tube containing 5 ml of LB broth with 100 μg/ml ampicillin. A tip without bacteria is used as the negative control. Clear solution in negative control the day after indicates sterile handling.
3. Incubate the culture in a shake incubator at 37 °C overnight.
4. Next day, transfer each of the 5 ml culture except for the negative control into a 1 l conical flask containing 250 ml LB broth with 100 μg/ml ampicillin.
5. Incubate the flasks in a shake incubator at 37 °C overnight.

#### 3.1.3 Plasmid DNA Extraction

1. Precipitate the bacteria by centrifuging the overnight culture at 6000 × *g* at 4 °C for 15 min.
2. Perform the maxi-prep reaction following the Qiagen Endo-Free Maxi kit protocol.
3. Measure the concentration of the plasmid DNA by UV spectrophotometer at 260 nm.
4. Save the plasmid DNA at −20 °C or proceed to the following step.

### 3.2 Hydrodynamic Tail Vein Injection

#### 3.2.1 Injection Solution Preparation

1. Mix the transgene plasmid and transposase plasmid in a 5:1 mass-to-mass ratio (30 μg of DNA in total per mouse) [18].
2. Measure the weight of each mouse on a scale.
3. Prepare the injection solution for each mouse by adding saline solution to the plasmid mix to a final volume that equals to 10% of the body weight of that mouse [13, 18].

   The volume of saline solution added is calculated using the following formula:

$$\begin{aligned}\text{Volume of saline (ml)} = {} & 10\%\ \text{mouse body weight (g)} \\ & - \text{volume of plasmid mix (ml)}\end{aligned}$$

4. Warm up the injection solution to 25–37 °C prior to injection.
5. Prior to injection, draw up injectable solution into suitable syringe then change needle to 27/29 g.
6. Hold the syringe upstraight and gently tap it with fingers to remove air bubbles. Leave the syringe aside.

#### 3.2.2 Animal Preparation and Injection

1. Transfer a mouse into the heat box at 28–30 °C. Wait a few minutes (≤10 min) for the tail veins to dilate.
2. Move the mouse from the heat box into a restrainer. Tighten the restrainer to hold mouse in position (Fig. 2).

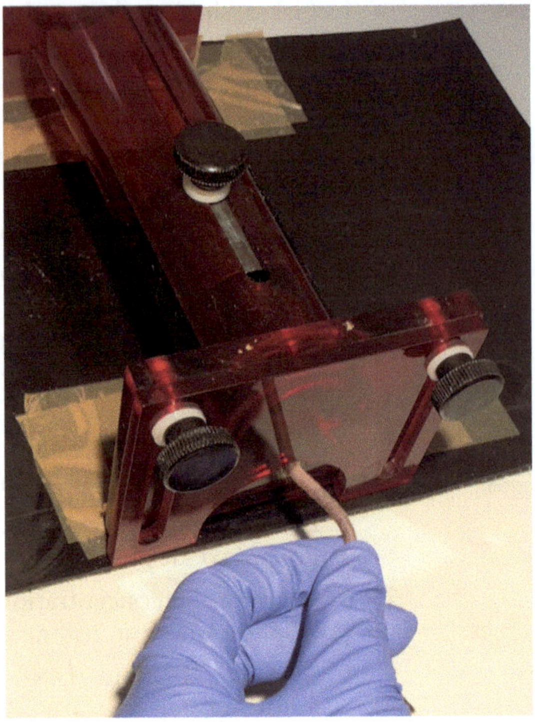

**Fig. 2** Mouse to be injected in a restrainer

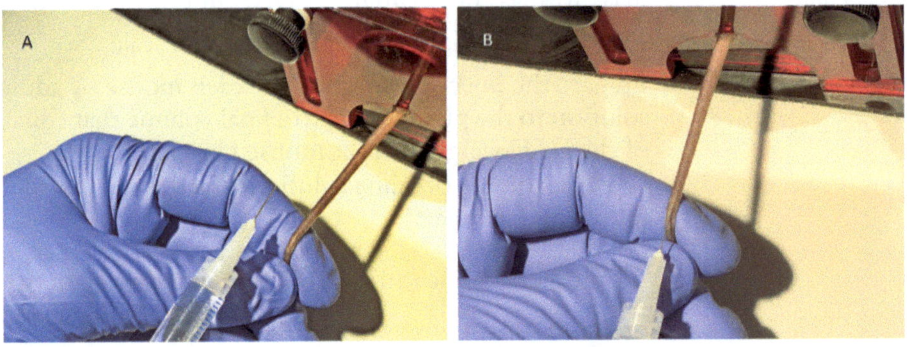


**Fig. 3** (**a**) Mouse tail held by forefinger and thumb and needle in bevel up position. (**b**) Position of the needle for injection

3. Hold the tail using the thumb and forefinger of nondominant hand to check the veins (Fig. 3a). One central artery and two lateral veins should be visible at the point.
4. Clean the tail with sterile saline on a swab. Gently apply EMLA cream to the spot of injection as local anesthesia.
5. Hold the tail under tension and gently insert the needle, bevel up, into one of the lateral veins (Fig. 3a). Move the needle roughly parallel to the tail and insert about 3 mm into the vein (*see* **Note 2**) (Fig. 3b).

6. A flash of blood will come into the needle hub indicating the successful insertion of the needle into the vein.
7. Inject suitable volume of injection solution into the vein within 10 s at a constant speed (*see* **Note 3**).
8. Remove the needle and apply gentle pressure to the vein using a swab for about 1 min to stop bleeding.
9. Loosen the restrainer and return the mouse to a holding cage. Monitor it for 5 min to ensure there is no further bleeding before returning it to the home cage.
10. Repeat **steps 1–9** until all mice are injected.
11. Closely monitor the recovery of mice and make sure there is no sign of distress (*see* **Note 4**).

### 3.3 Tumor Evaluation

#### 3.3.1 Liver Harvesting and Fixation

1. Maintain mice in standard Rat and Mouse No. 1 (RM1) diet.
2. Sacrifice the mice 71 days post injection or when termination conditions are fulfilled [11, 19].
3. Weigh each mouse on a scale before euthanasia.
4. Transfer a mouse to working surface and perform cervical dislocation.
5. Decapitate the mouse as the secondary method of euthanasia.
6. Sterilize the body surface of the mouse using 70% ethanol.
7. Open the abdomen with tweezers and scissors.
8. Take out the entire liver and measure the weight on a scale.
9. Count the number of nodules formed on the liver.
10. Evaluate the size of livers and check for GFP expression using a stereomicroscope.
11. Cut pieces ($\leq$15 mm in diameter) from nodules formed on the liver or from the big lobule for mice in control group.
12. Immediately transfer the tissue piece into a cassette and immerse the cassettes into 10% neutral buffered formalin (NBF). The volume of NBF should be at least 20 times volume of the liver to ensure efficient fixation [20]. Incubate at 4 °C overnight (*see* **Note 5**).
13. Next day, wash the tissue with 1× PBS for at least 30 min at 4 °C.
14. Dehydrate the tissues by immersing the cassettes into ascending concentrations of ethanol as following:

    Ethanol (70%), 2 washes, 1 h each;

    Ethanol (80%), 1 wash, 1 h;

    Ethanol (95%), 1 wash, 1 h;

    Ethanol (100%), 3 washes, 1.5 h each.

15. Immerse the cassettes into xylene for 1.5 h. Repeat this process twice to clear the ethanol in tissues.
16. Soak the tissues in paraffin wax suspension at 60 °C on a heat block for 2 h. Repeat this process once to ensure the removal of residual xylene.
17. Transfer the tissues to the embedding station.
18. Add a small amount of warm paraffin (60 °C) into the mold.
19. Place the tissue in the mold and adjust the orientation with tweezers.
20. Put the cassette on the top of the mold and fill the mold with paraffin.
21. Leave the paraffin blocks to harden on the cold plate.

### *3.3.2 Histopathological Analysis: Hematoxylin and Eosin (H&E) Staining*

1. Slice the paraffin-infiltrated tissues into 5 μm thick slices on a microtome.
2. Transfer the paraffin slices into a water bath at 40–45 °C.
3. Mount each slice into a microscopy slide and dry it at 40 °C overnight in an oven.
4. Next day, dewax and rehydrate the section by first warming up the slide in the oven for 15 s, followed by dipping the slides into a series of solvents in the following order:

   Xylene, 3 washes, 2 min each;

   Ethanol (100%), 2 washes, 2 min each;

   Ethanol (95%), 1 wash, 2 min;

   Ethanol (70%), 1 wash, 2 min;

   Ethanol (50%), 1 wash, 2 min;

   Tap water, 1 wash, 2 min.
5. Dye the nuclei with hematoxylin solution. Rinse immediately under running water for 2 min.
6. Bluing the slide in 1× Scotts tap water substitute for 30 s and quickly rinse the slide in tap water.
7. Differentiate the slide in 95% ethanol for 1 min.
8. Stain the cytoplasm and extracellular matrix in Eosin Y solution for 10 s.
9. Dehydrate and clear the slides with ethanol and xylene in the following order:

   Ethanol (95%), 2 washes, 1 min each;

   Ethanol (100%), 2 washes, 1 min each;

   Xylene, 3 washes, 1 min each.
10. Add mounting medium to a cover slide and slowly place the stained slide on the top, facing downward. Make the mounting

medium harden by leaving the slide at 40 °C overnight in the oven.

11. Examine the slides using light microscope. The nuclei should show blue-black color, while the cytoplasm and extracellular matrix are dyed pink. Check for changes of liver morphology sections from mice injected with the transgene plasmids and control plasmid. Typical micromorphological features for HCC include presence of macrotrabeculae (>3 cells wide), vascularization, and disintegrated reticulin network [21]. Increased-size hepatocytes with large nuclei, prominent nucleoli and infiltration of immune cells can also be observed [22].

#### 3.3.3 Immuno-histochemistry (IHC) Analysis

1. Prepare the slides and dewax the paraffin-bound tissues as in Subheading 3.3.2, **steps 1–4**.
2. Load the slides to a plastic slide rack and immerse the rack to a beaker containing citric acid buffer.
3. Microwave heat the buffer with high power for 10 min (95–100 °C).
4. Remove the beaker from microwave oven and leave the slides in hot buffer for 15 min. This is to reduce protein cross-linking and increase antibody binding [23].
5. Gently rinse the slides in running tap water for 5 min.
6. Quench the endogenous peroxidases activity by soaking the slides into 30% hydrogen peroxide for 15 min.
7. Gently rinse the slides in running tap water for 5 min.
8. Wash the slides with 1× PBS for 5 min.
9. Lay out slides in hybridization box and draw circle around the sample without touching the sample using a hydrophobic pen.
10. Create a humidified chamber in the hybridization box by laying water-soaked paper towels at the bottom.
11. Add 100–400 μl StartingBlock™ T20 (PBS) blocking buffer (*see* **Note 6**) onto the top of each slide. Incubate at room temperature for 15 min.
12. Remove the excessive blocking buffer with pipettes. Quickly rinse the slides with 1× PBS to remove the blocking reagent.
13. Dilute primary antibody in 1× BSA/PBT in an optimum ratio. Recommended working ratio is following (*see* **Note 7**):

    Anti-ki67 1:200.

    Glypican-3 1:200.

    Anti-p53 1:200.

    Anti-NRAS 1:500.

14. Add anti-ki67, anti-p53, and anti-NRAS primary antibodies to different sections of the tissue slide.
15. Incubate at 4 °C overnight in the humidified chamber.
16. Next day, wash the slides in 1× PBS twice with 5 min each.
17. Dilute secondary antibodies 1:200 in 1× BSA/PBT solution.
28. Apply secondary antibody onto the top of tissue section and incubate at 37 °C for 30 min.

    Anti-Ki67, anti-NRAS—anti-rabbit.
    Anti-Glypican3, anti-p53—anti-mouse.
19. Meanwhile, prepare HRP-ABC solution by mixing 200 μl solution A and 200 solution B from VECTASTAIN® Elite® ABC HRP Kit in 20 ml 1× BSA/PBT solution. Allow the mixture to stand for at least 30 min.
20. Wash the slides in 1× PBS twice with 5 min each.
21. Cover the top of the slide with HRP-ABC solution prepared in **step 23**. Incubate at 37 °C for 30 min.
22. Wash the slides in 1× PBS twice with 5 min each.
23. Develop the signal using DAB Peroxidase (HRP) Substrate Kit (with Nickel) following the manufacturer's protocol.
24. Soak the slides in $dH_2O$ for at least 5 min.
25. Counterstain the slides by quickly immersing slides into hematoxylin and immediately wash with $dH_2O$ (*see* **Note 8**).
26. Dehydrate and mount the slides following **steps 9** and **10** in Subheading 3.3.2.
27. Examine the slides using light microscope. The nuclei should show blue color from hematoxylin staining and the protein of interest should show brown color from the DAB staining.
28. Select 5–8 fields per slide and count the number of positive cells (brown-colored) using high magnification (×400). Calculate the labeling index using the following formula:

$$\text{Labelling index} = \frac{\text{Total number of positive cells}}{\text{Number of fields}} \div \text{Area of each field}$$

29. Statistical analysis: apply two-way ANOVA test to compare the labeling index between experimental and control groups. Statistical significance is considered when $p < 0.05$. HCC sections are expected to show higher level of Ki67 due to higher mitotic activity and higher cytoplasmic expression of Glypican-3 as a biomarker compared to mice in control group [24, 25]. Liver sections from mice transfected with plasmid containing *N-Ras*$^{G12V}$ are expected to have higher level of N-RAS compared with the control to show successful delivery of plasmids.

Liver sections from mice transfected with plasmid containing *shp53* are expected to have lower level of p53 compared with the control as an indication of successful delivery of plasmids.

## 4 Notes

1. If there is a lawn of colonies, reduce the volume of mixture used for spreading.
2. It is recommended to start the injection from the distal part of the tail, so that the higher part can be preserved for future injections in case the first one fails.
3. The vein will change color as the fluid passes through if the needle is in place correctly.
4. Typical signs of distress and/or pain for mice include orbital narrowing, nose and cheek bulge, ears moving backward and whisker position changes (forward or backward from the baseline) [26].
5. It is reported that the rate of penetration for aldehyde fixes is approximately 2–3 mm/h [27]. Insufficient fixing time may result in degradation of the interior tissues.
6. Alternatively, block using 5–10% normal serum from host species of the secondary antibody for 30 min at room temperature.
7. The optimum working concentration for each antibody may be different depending on the solvent, brand, lab environment, and so on. It is recommended to perform a titration of each antibody before using to obtain the optimised working concentration.
8. Quickly check the results under microscope. Try to achieve a good balance with DAB reaction and hematoxylin dye (e.g., dark brown/dark blue, light brown/light blue). If more blue color is needed, redip the slides into hematoxylin solution. If the blue color is too strong, the slides can be destained using 0.5% acid alcohol.

## References


1. Bray F, Ferlay J, Soerjomataram I et al (2018) Global cancer statistics 2018: GLOBOCAN estimates of incidence and mortality worldwide for 36 cancers in 185 countries. CA Cancer J Clin 68(6):394–424. https://doi.org/10.3322/caac.21492
2. Forner A, Reig M, Bruix J (2018) Hepatocellular carcinoma. Lancet 391 (10127):1301–1314. S0140-6736(18) 30010-2 [pii]
3. He L, Tian DA, Li PY et al (2015) Mouse models of liver cancer: progress and

recommendations. Oncotarget 6 (27):23306–23322. https://doi.org/10.18632/oncotarget.4202. eCollection

4. Suda T, Liu D (2007) Hydrodynamic gene delivery: its principles and applications. Mol Ther 15(12):2063–2069. S1525-0016(16)32719-8 [pii]
5. Duan M, Hao J, Cui S et al (2018) Diverse modes of clonal evolution in HBV-related hepatocellular carcinoma revealed by single-cell genome sequencing. Cell Res 28 (3):359–373. https://doi.org/10.1038/cr.2018.11
6. Carlson CM, Frandsen JL, Kirchhof N et al (2005) Somatic integration of an oncogene-harboring Sleeping Beauty transposon models liver tumor development in the mouse. Proc Natl Acad Sci U S A 102(47):17059–17064. https://doi.org/10.1073/pnas.0502974102
7. Yant SR, Meuse L, Chiu W et al (2000) Somatic integration and long-term transgene expression in normal and haemophilic mice using a DNA transposon system. Nat Genet 25(1):35–41. https://doi.org/10.1038/75568
8. Ivics Z, Hackett PB, Plasterk RH et al (1997) Molecular reconstruction of Sleeping Beauty, a Tc1-like transposon from fish, and its transposition in human cells. Cell 91(4):501–510. S0092-8674(00)80436-5 [pii]
9. Yant SR, Park J, Huang Y et al (2004) Mutational analysis of the N-terminal DNA-binding domain of Sleeping Beauty transposase: critical residues for DNA binding and hyperactivity in mammalian cells. Mol Cell Biol 24 (20):9239–9247. https://doi.org/10.1128/MCB.24.20.9239-9247.2004
10. Heindryckx F, Colle I, Van Vlierberghe H (2009) Experimental mouse models for hepatocellular carcinoma research. Int J Exp Pathol 90(4):367–386. https://doi.org/10.1111/j.1365-2613.2009.00656.x
11. Keng VW, Tschida BR, Bell JB et al (2011) Modeling hepatitis B virus X-induced hepatocellular carcinoma in mice with the Sleeping Beauty transposon system. Hepatology 53 (3):781–790. https://doi.org/10.1002/hep.24091
12. Chung SI, Moon H, Ju HL et al (2016) Comparison of liver oncogenic potential among human RAS isoforms. Oncotarget 7 (6):7354–7366. https://doi.org/10.18632/oncotarget.6931. eCollection
13. Kang TW, Yevsa T, Woller N et al (2011) Senescence surveillance of pre-malignant hepatocytes limits liver cancer development. Nature 479(7374):547–551. https://doi.org/10.1038/nature10599
14. Bieging KT, Mello SS, Attardi LD (2014) Unravelling mechanisms of p53-mediated tumour suppression. Nat Rev Cancer 14 (5):359–370. https://doi.org/10.1038/nrc3711
15. Teramoto T, Satonaka K, Kitazawa S et al (1994) P53 gene abnormalities are closely related to hepatoviral infections and occur at a late stage of hepatocarcinogenesis. Cancer Res 54(1):231–235
16. Schulze K, Imbeaud S, Letouze E et al (2015) Exome sequencing of hepatocellular carcinomas identifies new mutational signatures and potential therapeutic targets. Nat Genet 47 (5):505–511. https://doi.org/10.1038/ng.3252
17. Wiesner SM, Decker SA, Larson JD et al (2009) De novo induction of genetically engineered brain tumors in mice using plasmid DNA. Cancer Res 69(2):431–439. https://doi.org/10.1158/0008-5472.CAN-08-1800
18. Rudalska R, Dauch D, Longerich T et al (2014) In vivo RNAi screening identifies a mechanism of sorafenib resistance in liver cancer. Nat Med 20(10):1138–1146. https://doi.org/10.1038/nm.3679
19. Tordella L, Khan S, Hohmeyer A et al (2016) SWI/SNF regulates a transcriptional program that induces senescence to prevent liver cancer. Genes Dev 30(19):2187–2198. gad.286112.116 [pii]
20. Bancroft JD, Gamble M (2008) Theory and practice of histological techniques, 6th edn. Churchill Livingstone, Edinburgh
21. Schlageter M, Terracciano LM, D'Angelo S et al (2014) Histopathology of hepatocellular carcinoma. World J Gastroenterol 20 (43):15955–15964. https://doi.org/10.3748/wjg.v20.i43.15955
22. Naas T, Ghorbani M, Alvarez-Maya I et al (2005) Characterization of liver histopathology in a transgenic mouse model expressing genotype 1a hepatitis C virus core and envelope proteins 1 and 2. J Gen Virol 86 (Pt 8):2185–2196. 86/8/2185 [pii]
23. Brown RW, Chirala R (1995) Utility of microwave-citrate antigen retrieval in diagnostic immunohistochemistry. Mod Pathol 8 (5):515–520

24. Shafizadeh N, Ferrell LD, Kakar S (2008) Utility and limitations of glypican-3 expression for the diagnosis of hepatocellular carcinoma at both ends of the differentiation spectrum. Mod Pathol 21(8):1011–1018. https://doi.org/10.1038/modpathol.2008.85

25. Luo Y, Ren F, Liu Y et al (2015) Clinicopathological and prognostic significance of high Ki-67 labeling index in hepatocellular carcinoma patients: a meta-analysis. Int J Clin Exp Med 8(7):10235–10247

26. Langford DJ, Bailey AL, Chanda ML et al (2010) Coding of facial expressions of pain in the laboratory mouse. Nat Methods 7 (6):447–449. https://doi.org/10.1038/nmeth.1455

27. Paljarvi L, Garcia JH, Kalimo H (1979) The efficiency of aldehyde fixation for electron microscopy: stabilization of rat brain tissue to withstand osmotic stress. Histochem J 11 (3):267–276. https://doi.org/10.1007/bf01005026

# Chapter 15

# Mouse Model of Alcoholic Steatohepatitis


## Adeline Bertola



## Abstract

Alcoholic liver disease (ALD) is one of the most common causes of chronic liver disease in Western countries. The spectrum of ALD ranges from simple steatosis to steatohepatitis to cirrhosis and hepatocellular carcinoma. Over the past 50 years, several animal models of ALD have been developed. Although none of them faithfully recapitulates the human disease, they have proven to be invaluable tools to study the pathogenesis of ALD, to identify potential therapeutic targets and to test new drugs. Here, we describe the mouse model of chronic and binge ethanol feeding, also known as the NIAAA model or Gao binge model. This model combines chronic feeding of Lieber–DeCarli ethanol liquid diet with acute administration of high-dose ethanol by oral gavage to mimic the drinking patterns of many alcoholic patients who engage in episodes of binge drinking on top of chronic daily drinking. Short-term (10-day) chronic plus single binge ethanol feeding causes a substantial increase in serum transaminase levels, moderate steatosis and mild inflammation characterized by lobular neutrophil infiltration. Long-term (8-week) chronic plus single or multiple (twice a week) binge ethanol feeding induce more severe steatohepatitis and mild fibrosis. This clinically relevant, easy-to-perform model of ALD is currently used by many research laboratories to reproduce early stages of human alcoholic steatohepatitis.



**Key words** Mouse model, Alcoholic liver disease, Fatty liver, Steatohepatitis, Fibrosis, Chronic and binge ethanol feeding


## 1 Introduction

Excessive alcohol consumption is the most prevalent cause of advanced liver disease in Western countries [1]. Alcoholic liver disease (ALD) encompasses a broad spectrum of hepatic disorders ranging from alcoholic fatty liver to more severe forms of liver injury including alcoholic steatohepatitis (ASH), alcoholic cirrhosis, and superimposed hepatocellular carcinoma. Fatty liver is the earliest response of the liver to excessive alcohol intake and develops in more than 90% of heavy drinkers. It is usually considered benign and resolves with abstinence. ASH occurs in up to 35% of alcoholic patients and is characterized by the presence of hepatic steatosis accompanied by hepatocellular injury and lobular inflammation consisting mainly of neutrophilic infiltrates. ASH is often associated


Myriam Aouadi and Valerio Azzimato (eds.), *Kupffer Cells: Methods and Protocols*, Methods in Molecular Biology, vol. 2164, https://doi.org/10.1007/978-1-0716-0704-6_15,

with progressive fibrosis which can lead to cirrhosis and end-stage liver disease. In addition, patients with underlying ASH or alcoholic cirrhosis and recent heavy alcohol drinking can also develop alcoholic hepatitis, a form of acute-on-chronic liver disease associated with high mortality [2, 3]. Despite the burden of ALD, its pathogenesis is still poorly understood and no targeted therapy has been approved. Studies have been hampered, in part, by the lack of readily available experimental models of advanced ALD. Rodents have a natural aversion to alcohol and metabolize alcohol at faster rate than humans, making it challenging to maintain blood alcohol concentration high enough to cause significant liver damage.

Over the last five decades, three major mouse models of ALD have been developed.

In the 1960s, Dr. Lieber's group developed a "forced choice" model of alcohol feeding in which ethanol is incorporated into a nutritionally adequate liquid diet (referred to as the Lieber–DeCarli liquid diet) provided as the only source of food and water, thereby partially overcoming the rodents' aversion to alcohol. Chronic ad libitum feeding with the Lieber–DeCarli liquid diet containing ethanol for up to 8 weeks has been widely used. However, in most studies, the liver pathology is predominantly limited to steatosis [4, 5]. "Second hits" such as exposure to lipopolysaccharide or carbon tetrachloride have been used to induce steatohepatitis and liver fibrosis in chronic ethanol-fed mice but the relevance of these models to human ALD has been questioned [6].

In the 1980s, Drs. French and Tsukamoto developed the intragastric ethanol infusion model in which ethanol is administered together with a liquid diet directly to the stomach using a surgically implanted gastrostomy catheter. When combined with ad libitum high-cholesterol, high-fat diet feeding, intragastric ethanol infusion induces severe steatohepatitis and mild fibrosis within 8 weeks [7]. However, the use of this model has been limited due its high technical demand and the requirement of extensive care and monitoring of animals [8].

In the 2010s, Dr. Gao's group developed the mouse model of chronic and binge ethanol feeding, also known as the NIAAA model or Gao binge model, in which binge exposure to high dose of ethanol itself serves as the "second hit" that causes progression from steatosis to steatohepatitis in chronically ethanol-fed mice. In the original chronic and binge ethanol feeding model, mice are initially fed ad libitum with the control Lieber–DeCarli liquid diet for 5 days to acclimatize them to liquid diet feeding. Ethanol-fed mice are then fed ad libitum for 10 days with the ethanol Lieber–DeCarli liquid diet containing 5% (vol/vol) ethanol (i.e., 27.6% of total calories derived from ethanol) and control mice are pair-fed with the control Lieber–DeCarli liquid diet. In the early morning of the 11th day, ethanol-fed mice are administered a single dose of ethanol (5 g/kg) by oral gavage and control mice are gavaged with an isocaloric maltodextrin solution (9 g/kg). Mice are euthanized 9 h after gavage (Fig. 1). This short-term chronic and single binge ethanol feeding causes a substantial increase in serum alanine aminotransferase (ALT) and aspartate aminotransferase (AST) levels, moderate steatosis and mild inflammation characterized by lobular neutrophil infiltration [9–13]. Long-term (chronic) plus binge ethanol feeding model has also been developed, in which the chronic administration of ethanol is extended to 8 weeks. Mice are gavaged with a single dose of ethanol at the end of the 8-week chronic feeding period (long-term chronic plus single binge ethanol feeding) or gavaged twice a week throughout the 8-week chronic feeding period (long-term chronic plus multiple binge ethanol feeding) (Fig. 2). These long-term chronic and binge ethanol feeding models induce more severe steatohepatitis and mild chicken-wire fibrosis [14]. Collectively, the chronic and binge ethanol feeding model is easy to perform and closely approximates the drinking patterns observed in many alcoholic patients and reproduces early stages of human ASH. The mechanisms

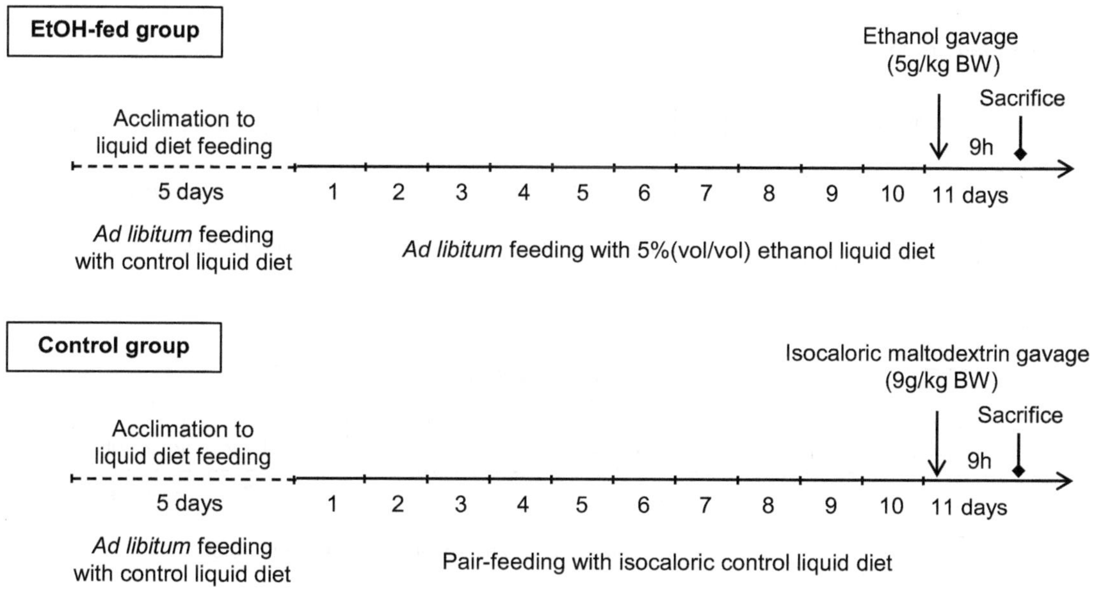


**Fig. 1** Overview of the chronic and binge ethanol feeding model. The chronic and binge ethanol feeding protocol comprises three stages. *Stage 1. Acclimation to liquid diet feeding.* Mice are fed ad libitum with the control Lieber–DeCarli liquid diet for 5 days. *Stage 2. Chronic administration of ethanol and pair-feeding.* Ethanol-fed mice are fed ad libitum with the ethanol Lieber–DeCarli liquid diet containing 5% (vol/vol) ethanol for 10 days and the control mice are pair-fed with the isocaloric control Lieber–DeCarli liquid diet. The control and ethanol liquid diets are the only source of food and water; they are changed daily in the late afternoon. *Stage 3. Acute administration of ethanol and maltodextrin by oral gavage.* In the early morning of the 11th day, ethanol-fed mice are administered a single dose of ethanol (5 g/kg) by oral gavage and control mice are gavaged with an isocaloric maltodextrin solution (9 g/kg). Mice are euthanized 9 h after gavage

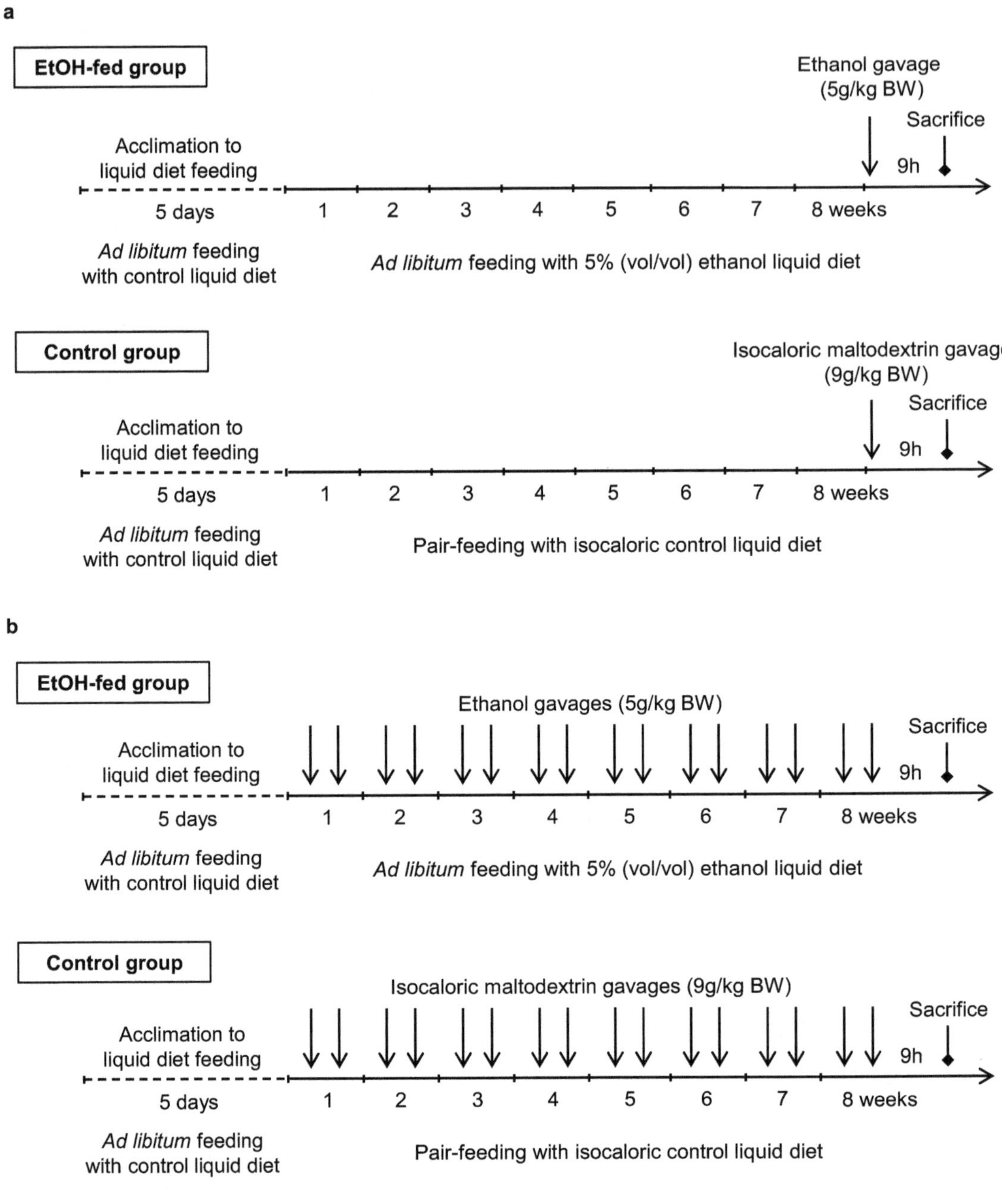


**Fig. 2** Overview of the long-term chronic and binge ethanol feeding models. In the long-term chronic and binge ethanol feeding models, the stage of chronic administration of ethanol and pair-feeding is extended to 8 weeks. (**a**) In the long-term chronic plus single binge ethanol feeding, ethanol- and pair-fed mice are gavaged at the end of the 8-week chronic feeding period with a single dose of ethanol or isocaloric maltodextrin, respectively. Mice are euthanized 9 h after gavage. (**b**) In the long-term chronic plus multiple binge ethanol feeding, ethanol- and pair-fed mice are gavaged twice a week throughout the 8-week chronic feeding period with ethanol or isocaloric maltodextrin, respectively. Mice are euthanized 9 h after the last gavage

underlying chronic and binge ethanol feeding-induced liver injury have recently been reviewed [15].

In this chapter, we describe how to implement the chronic and binge ethanol feeding protocol and discuss the factors that affect alcohol-induced liver injury in this model.

## 2 Materials

### 2.1 Animals and Reagents

1. Animals: Sex- and age-matched C57BL/6 mice (*see* **Note 1**).
2. Rodent Liquid Diet, Lieber–DeCarli '82, Shake and Pour, Control, 4 l/Bag (Product# F1259SP, Bio-Serv).
3. Rodent Liquid Diet, Lieber–DeCarli '82, Shake and Pour, Ethanol, 4 l/Bag (Product# F1258SP, Bio-Serv).

   The caloric profile of these diets is presented in Table 1. Their nutritional profile is available online (http://www.bio-serv.com/Rodent_Liquid_Diets/LD82SP.html) (*see* **Note 2**).
4. Maltodextrin, 10DE, Food Grade (Product# 3585, Bio-Serv).
5. Ethanol, 95% (vol/vol) (*see* **Note 3**).
6. Mouse drinking water.

### 2.2 Equipment

1. Laboratory scale.
2. Commercial food blender.
3. Storage bottles.
4. Measuring cylinder.
5. Serological pipets and pipetting device.
6. Liquid diet feeding tubes with screw caps (50 ml size, Product# 9019, Bio-Serv) (*see* **Note 4**).
7. Tube holders (Product# 9015, Bio-Serv).
8. Wood gnawing blocks (*see* **Note 5**).
9. Oral gavage needle (*see* **Note 6**).

**Table 1**
**Caloric profile of the control and ethanol liquid diets**

| | F1259SP—Control (kcal/l) | F1258SP—Ethanol (kcal/l) |
|---|---|---|
| Protein | 151 | 151 |
| Fat | 359 | 359 |
| Carbohydrate | 490 | 135 |
| Ethanol + Maltodextrin | 0 | 355 |
| **Total** | **1000** | **1000** |

10. Syringe, 1 ml.
11. Heat therapy pump and reusable pad (Adroit HTP-1500).

### 2.3 Preparation of Diets

The directions for preparing 1 l of control and ethanol liquid diets are described below. Scale up or down as needed.

#### 2.3.1 Control Liquid Diet

1. Weigh 226 g of the dry mix (Product# F1259SP, Bio-Serv).
2. Measure 860 ml of water.
3. Add the dry mix and water into the blender and mix until no clumps remain. Refrigerate the prepared control liquid diet and use within 3 days to prevent diet deterioration.

#### 2.3.2 Ethanol Liquid Diet

To achieve the desired percentage of ethanol in the diet, use Table 2 to determine the corresponding amounts of maltodextrin, water and 95% ethanol to add to the dry mix. The resulting diet is isocaloric to the control diet.

1. Weigh 133 g of the dry mix (Product# F1258SP, Bio-Serv).
2. Weigh the maltodextrin (19.7 g for a 5% (vol/vol) ethanol diet) (Product# 3585, Bio-Serv).
3. Measure the water (910 ml for a 5% (vol/vol) ethanol diet).
4. Add the dry mix, maltodextrin and water into the blender and mix until no clumps remain. Refrigerate the prepared liquid diet (without ethanol) and use within 3 days to prevent diet deterioration.
5. Prior to feeding, add the 95% ethanol (52.6 ml of 95% ethanol for a 5% (vol/vol) ethanol diet) and mix thoroughly (*see* **Note 7**).

**Table 2**
**Mixing directions to prepare 1 l of ethanol liquid diet**

| Ethanol content in the diet | | | Add | | | |
|---|---|---|---|---|---|---|
| % Ethanol (vol/vol) | % Ethanol (wt/vol) | % Ethanol derived calories | Dry mix (g) | Maltodextrin (g) | Water (ml) | 95% Ethanol (ml) |
| 1 | 0.8 | 5.5 | 133 | 74.9 | 900 | 10.5 |
| 2 | 1.6 | 11.0 | 133 | 61.1 | 910 | 21.1 |
| 3 | 2.4 | 16.6 | 133 | 47.3 | 910 | 31.6 |
| 4 | 3.2 | 22.1 | 133 | 33.5 | 920 | 42.1 |
| 5 | 3.9 | 27.6 | 133 | 19.7 | 910 | 52.6 |

Ethanol density = 0.789 g/ml
1 g ethanol = 7 kcal
1 g maltodextrin = 4 kcal

### 2.4 Preparation of Gavage Solutions

The directions for preparing 20 ml of gavage solutions are described below. Scale up or down as needed.

#### 2.4.1 Maltodextrin Gavage Solution

Prepare a 45% (wt/vol) maltodextrin solution by dissolving 9 g of maltodextrin in a final volume of 20 ml of water. Use immediately after preparation.

#### 2.4.2 Ethanol Gavage Solution

Prepare a 31.5% (vol/vol) ethanol solution by mixing 6.6 ml of 95% ethanol and 13.4 ml of water. Use immediately after preparation.

## 3 Methods

### 3.1 Acclimation to Liquid Diet Feeding (Ad Libitum Feeding with Control Liquid Diet for 5 Days)

1. Prepare an adequate number of clean cages (containing bedding, nesting material and a wood gnawing block), feeding tubes with screw caps, and tube holders according to your experimental groups. Mice are housed singly or in pairs during the feeding.
2. Prepare an adequate volume of control liquid diet. You will need 50 ml of diet per cage per day.
3. Dispense the control liquid diet into the feeding tubes. Place your gloved thumb over the feed well of the tube and fill up to 50 ml (Fig. 3a). Maintain your thumb on the feed well and seal the feeding tube by screwing the cap on tightly.
4. Place the feeding tubes containing the control liquid diet in the clean cages using the tube holders (Fig. 3b) (*see* **Notes 8** and **9**).
5. Transfer the mice to the cages.

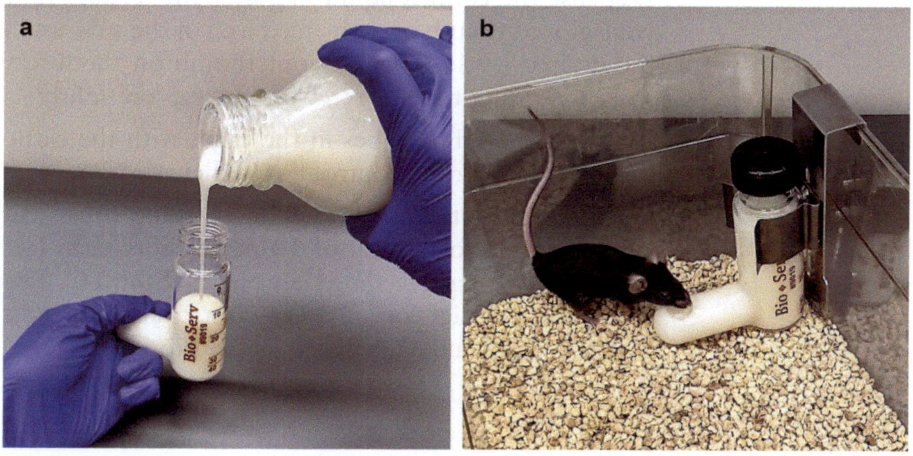


**Fig. 3** Feeding tube setup. (**a**) The control and ethanol Lieber–DeCarli liquid diets are administered in graduated feeding tubes specifically designed to dispense liquid diets. (**b**) Feeding tubes are positioned in the cages using holders designed to fit over the lip of the cage

6. Allocate the cages to control and ethanol-fed groups.
7. Change the control liquid diet and feeding tubes daily in the late afternoon for a total of 5 days (*see* **Note 10**).

If you are using mice on a genetic background other than C57BL/6, *see* **Note 11**.

### 3.2 Chronic Ethanol Feeding and Pair-Feeding

#### 3.2.1 Ad Libitum Feeding of Ethanol-Fed Mice with 5% (Vol/Vol) Ethanol Liquid Diet

1. Prepare an adequate number of feeding tubes with screw caps and an adequate volume of 5% (vol/vol) ethanol liquid diet (*see* **Note** 7). You will need 50 ml of diet per cage per day.
2. Dispense the 5% (vol/vol) ethanol liquid diet into the feeding tubes. Place your gloved thumb over the feed well of the tube and fill up to 50 ml. Maintain your thumb on the feed well and seal the feeding tube by screwing the cap on tightly.
3. Replace the feeding tubes in the cages with the clean feeding tubes containing the 5% (vol/vol) ethanol liquid diet (*see* **Note 8**).
4. Change the ethanol liquid diet and feeding tubes daily in the late afternoon for a total of 10 days (short-term feeding) or 8 weeks (long-term feeding) (*see* **Notes 9** and **10**).
5. Record the food intake for each cage on each day and calculate the average volume of ethanol diet consumed per mouse per day.

#### 3.2.2 Pair-Feeding of Control Mice with Control Liquid Diet

1. Prepare an adequate number of feeding tubes with screw caps and an adequate volume of control liquid diet. Match the amount of control liquid diet provided to the control mice to the average volume of ethanol diet consumed by the ethanol-fed mice on the previous day (*see* **Note 12**).
2. Dispense the control liquid diet into the feeding tubes. Place your gloved thumb over the feed well of the tube and fill up to the desired level. Maintain your thumb on the feed well and seal the feeding tube by screwing the cap on tightly.
3. Replace the feeding tubes in the cages with the clean feeding tubes containing the control liquid diet (*see* **Note 8**).
4. Change the control liquid diet and feeding tubes daily in the late afternoon for a total of 10 days (short-term feeding) or 8 weeks (long-term feeding) (*see* **Notes 9** and **10**). Ensure that the amount of diet consumed by control and ethanol-fed mice is equalized on a day-to-day basis (*see* **Note 13**).

### 3.3 Oral Gavage of the Ethanol and Maltodextrin Solutions

#### 3.3.1 Short-Term and Long-Term Chronic Plus Single Binge Ethanol Feeding

1. In the early morning of the following day, prepare an adequate volume of 31.5% (vol/vol) ethanol and 45% (wt/vol) maltodextrin solutions (*see* **Notes 14** and **15**). The volume in microliters of ethanol and maltodextrin solutions to be administered is 20 times the mouse body weight in grams (e.g., 400 μl for a mouse weighting 20 g).
2. Weigh the mice and calculate the volume of solution to be administered.
3. Select an appropriate gavage needle for the size of the mice by measuring the distance from the corner of the mouth to the xiphoid process (Fig. 4a).
4. Attach the gavage needle to a 1 ml syringe and fill the syringe with the desired volume of gavage solution.
5. Gently but firmly restrain the mice in an upright position. Grab the mice by the scruff of the neck with your thumb and forefinger and extend the grasp down the animal's back using your middle and ring fingers so that the fore legs are extended out to the side, the head and neck are immobilized and the spine is straight (Fig. 4b). Ensure the animal can breathe freely.
6. Insert the gavage needle into the mouth through the diastema between the incisors and the cheek teeth.
7. Slide the gavage needle along the roof of the mouth and extend the head and neck back using the needle as a lever to align the neck and esophagus.

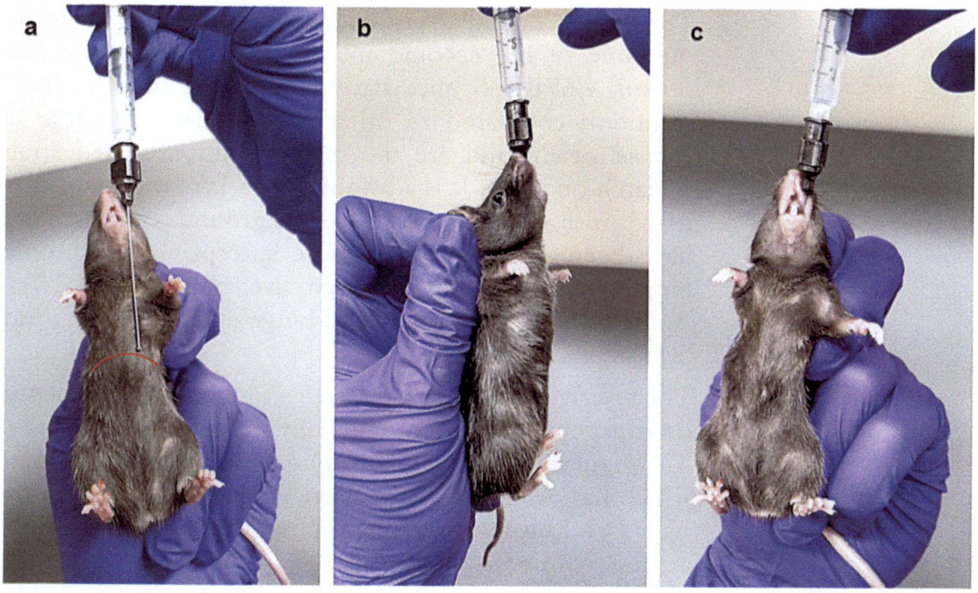


**Fig. 4** Oral gavage. (**a**) Measurement of the length of the gavage needle to be inserted into the esophagus. (**b, c**) Administration of the gavage solution upon proper restraint of the animal **(b)** and placement of the gavage needle **(c)**

8. Smoothly advance the needle through the esophagus (*see* **Note 16**).
9. Once the needle is properly positioned, slowly deliver the solution into the stomach (Fig. 4c) (*see* **Note 17**).
10. After dosing, smoothly remove the gavage needle following the same angle as insertion.
11. Return the mice to their original cages and place the cages on the heating pad (set point temperature of 37 °C) (*see* **Note 18**).
12. Monitor the mice for at least 10 min for any signs of distress following the oral gavage and continue monitoring the mice during their recovery from alcohol intoxication (*see* **Notes 19** and **20**).
13. Nine hours after gavage, weigh the mice and collect the blood (*see* **Note 21**).
14. Euthanize the mice and harvest and weigh the liver.
15. Process the liver for further analysis (*see* **Note 22**).

*3.3.2 Long-Term Chronic Plus Multiple Binge Ethanol Feeding*

Perform the oral gavage of the ethanol and maltodextrin solutions twice a week during the 8 weeks of chronic ethanol feeding and pair-feeding as described in the Subheading 3.3.1. Euthanize the mice 9 h after the last gavage.

## 4 Notes

1. The strain, sex and age of mice are critical factors affecting the outcome of the chronic and binge ethanol feeding model. *Strain*. C57BL/6 mice have a high voluntary alcohol consumption compared to other commercially available inbred mouse strains and are the most commonly used strain for research on ALD. It is advisable to backcross mutant or transgenic mice onto a C57BL/6 background. *Sex*. Female mice have been reported to be more susceptible to chronic and binge ethanol feeding-induced liver injury [11, 16]. *Age*. Aging exacerbates chronic and binge ethanol feeding-induced liver injury and fibrosis [17].
2. The severity of alcohol-induced liver injury is highly dependent on the diet composition, especially the amount and type of fat. The regular Lieber–DeCarli liquid diet is rich in unsaturated fat which has deleterious effects on alcohol-induced liver pathology [18]. Low fat and high protein Lieber–DeCarli liquid diets are also commercially available.
3. As 100% ethanol absorbs water from air, it is advisable to prepare a 95% (vol/vol) ethanol solution after opening a new bottle of 100% ethanol.

4. It is advisable to use feeding tubes specifically designed to dispense liquid diets to avoid clogging that occurs with sipper tubes.
5. Providing gnawing enrichment is essential to prevent dental malocclusion when mice are maintained on liquid diets.
6. Recommended gavage needle sizes are: 22 gauge, 1–1.5 in. for 15–20 g mice; 20 gauge, 1–2 in. for 20–25 g mice; 18 gauge, 1–2 in. for 25–30 g mice.
7. Add the 95% ethanol to the liquid diet just before dispensing to the mice. Do not store liquid diet containing ethanol.
8. The liquid diet is both a source of food and water. Do not provide a separate source of water.
9. Adjust the height of the feeding tube low enough to allow easy access to the liquid diet but high enough to avoid clogging with bedding material.
10. The time of the day when the liquid diet is changed has to be kept consistent throughout the feeding period. As mice consume the majority of their food during the dark phase, it is advisable to dispense the fresh liquid diet in the late afternoon, just before the onset of the dark phase. Cages should be checked again in the early morning to ensure that the feeding tubes are not clogged or leaking.
11. C57BL/6 mice are a high alcohol preferring strain and tolerate well the switch from the control diet to the 5% (vol/vol) ethanol diet. For other strains, it is advisable to gradually increase the concentrations of ethanol during the phase of acclimation to liquid diet feeding from 1% (vol/vol) on day 2 to 4% (vol/vol) on day 5. In that case, control mice are pair-fed with the control diet.
12. The volume of control liquid diet to be dispensed on the first day of pair-feeding is estimated based on previous experiments or a pilot study. Take into account the dead volume of the feeding tube. C57BL/6 mice generally consume an average of 10 ml of ethanol diet per day. Alternatively, the feeding start date of the ethanol and pair-fed groups can be staggered on two consecutive days.
13. Because of their aversion to alcohol, mice consume the ethanol diet in significantly lesser amounts than the control diet. Therefore, to study the effects of alcohol independently of changes in caloric intake, it is necessary to restrict the amount of control diet provided to the control mice to match the food intake of the ethanol-fed mice. It is advisable to weigh the mice prior to start of study and three times per week throughout the feeding period. Ethanol and pair-fed mice should have comparable body weight gain throughout the feeding period [19].

14. It is critical to perform the oral gavage in the early morning, just after the onset of the light phase. As mice consume the majority of the ethanol diet during the dark, blood alcohol concentration is elevated at the onset of the light phase and is further increased after ethanol gavage [11, 19].
15. Prepare the ethanol and maltodextrin gavage solutions just before administering to the mice. Do not store gavage solutions.
16. The mouse usually swallows as the gavage needle approaches the pharynx, allowing the needle to pass freely into the esophagus. If there is any resistance, do not force the needle down the esophagus. Smoothly remove the needle and try again. Do not perform oral gavage in anesthetized mice.
17. If the gavage solution is coming back up the esophagus into the mouth, immediately remove the gavage needle and release the mouse. Monitor for signs of labored breathing or respiratory distress due to aspiration of solution into the lungs.
18. It is critical to provide thermal support to minimize body heat loss and prevent hypothermia induced by ethanol.
19. Mice showing signs of labored breathing or respiratory distress following oral gavage should be euthanized.
20. Within minutes after ethanol gavage, mice will exhibit signs of heavy intoxication such as staggering gait and lack of muscular coordination, followed by lethargy and loss of consciousness, which dissipate as they recover within 5–6 h.
21. Time of sacrifice is critical. Serum concentrations of ALT and AST significantly increase 6 h postgavage, reach peak levels 9 h postgavage, and return to basal levels by 24 h [11, 19].
22. Fix a portion of the left lateral lobe in 10% neutral buffered formalin and process for histology (hematoxylin and eosin) and immunohistochemistry. Freeze the remaining liver tissue in liquid nitrogen and store at −80 °C for further analysis, including Oil Red O staining for evaluating steatosis, lipid extraction for measuring triglyceride levels, real-time PCR for studying gene expression, and western blot for protein analysis.

## References

1. Mathurin P, Bataller R (2015) Trends in the management and burden of alcoholic liver disease. J Hepatol 62(1 Suppl):S38–S46. https://doi.org/10.1016/j.jhep.2015.03.006
2. Gao B, Bataller R (2011) Alcoholic liver disease: pathogenesis and new therapeutic targets. Gastroenterology 141(5):1572–1585. https://doi.org/10.1053/j.gastro.2011.09.002
3. Mandrekar P, Bataller R, Tsukamoto H, Gao B (2016) Alcoholic hepatitis: translational approaches to develop targeted therapies. Hepatology 64(4):1343–1355. https://doi.org/10.1002/hep.28530
4. Bertola A (2018) Rodent models of fatty liver diseases. Liver Res 2(1):1–11. https://doi.org/10.1016/j.livres.2018.03.001

5. Mathews S, Xu M, Wang H, Bertola A, Gao B (2014) Animals models of gastrointestinal and liver diseases. Animal models of alcohol-induced liver disease: pathophysiology, translational relevance, and challenges. Am J Physiol Gastrointest Liver Physiol 306(10): G819–G823. https://doi.org/10.1152/ajpgi.00041.2014
6. Tsukamoto H, Machida K, Dynnyk A, Mkrtchyan H (2009) "Second hit" models of alcoholic liver disease. Semin Liver Dis 29 (2):178–187. https://doi.org/10.1055/s-0029-1214373
7. Lazaro R, Wu R, Lee S, Zhu NL, Chen CL, French SW, Xu J, Machida K, Tsukamoto H (2014) Osteopontin deficiency does not prevent but promotes alcoholic neutrophilic hepatitis in mice. Hepatology 61:129. https://doi.org/10.1002/hep.27383
8. Ueno A, Lazaro R, Wang PY, Higashiyama R, Machida K, Tsukamoto H (2012) Mouse intragastric infusion (iG) model. Nat Protoc 7 (4):771–781. https://doi.org/10.1038/nprot.2012.014
9. Bertola A, Park O, Gao B (2013) Chronic plus binge ethanol feeding synergistically induces neutrophil infiltration and liver injury in mice: a critical role for E-selectin. Hepatology 58 (5):1814–1823. https://doi.org/10.1002/hep.26419
10. Cai Y, Xu MJ, Koritzinsky EH, Zhou Z, Wang W, Cao H, Yuen PS, Ross RA, Star RA, Liangpunsakul S, Gao B (2017) Mitochondrial DNA-enriched microparticles promote acute-on-chronic alcoholic neutrophilia and hepatotoxicity. JCI Insight 2(14):e92634. https://doi.org/10.1172/jci.insight.92634
11. Ki SH, Park O, Zheng M, Morales-Ibanez O, Kolls JK, Bataller R, Gao B (2010) Interleukin-22 treatment ameliorates alcoholic liver injury in a murine model of chronic-binge ethanol feeding: role of signal transducer and activator of transcription 3. Hepatology 52 (4):1291–1300. https://doi.org/10.1002/hep.23837
12. Li M, He Y, Zhou Z, Ramirez T, Gao Y, Gao Y, Ross RA, Cao H, Cai Y, Xu M, Feng D, Zhang P, Liangpunsakul S, Gao B (2017) MicroRNA-223 ameliorates alcoholic liver injury by inhibiting the IL-6-p47(phox)-oxidative stress pathway in neutrophils. Gut 66 (4):705–715. https://doi.org/10.1136/gutjnl-2016-311861
13. Mathews S, Feng D, Maricic I, Ju C, Kumar V, Gao B (2016) Invariant natural killer T cells contribute to chronic-plus-binge ethanol-mediated liver injury by promoting hepatic neutrophil infiltration. Cell Mol Immunol 13 (2):206–216. https://doi.org/10.1038/cmi.2015.06
14. Xu MJ, Cai Y, Wang H, Altamirano J, Chang B, Bertola A, Odena G, Lu J, Tanaka N, Matsusue K, Matsubara T, Mukhopadhyay P, Kimura S, Pacher P, Gonzalez FJ, Bataller R, Gao B (2015) Fat-specific protein 27/CIDEC promotes development of alcoholic steatohepatitis in mice and humans. Gastroenterology 149(4):1030–1041.e1036. https://doi.org/10.1053/j.gastro.2015.06.009
15. Gao B, Xu MJ, Bertola A, Wang H, Zhou Z, Liangpunsakul S (2017) Animal models of alcoholic liver disease: pathogenesis and clinical relevance. Gene Expr 17(3):173–186. https://doi.org/10.3727/105221617X695519
16. Fulham MA, Mandrekar P (2016) Sexual dimorphism in alcohol induced adipose inflammation relates to liver injury. PLoS One 11 (10):e0164225. https://doi.org/10.1371/journal.pone.0164225
17. Ramirez T, Li YM, Yin S, Xu MJ, Feng D, Zhou Z, Zang M, Mukhopadhyay P, Varga ZV, Pacher P, Gao B, Wang H (2017) Aging aggravates alcoholic liver injury and fibrosis in mice by downregulating sirtuin 1 expression. J Hepatol 66(3):601–609. https://doi.org/10.1016/j.jhep.2016.11.004
18. Kirpich IA, Miller ME, Cave MC, Joshi-Barve-S, McClain CJ (2016) Alcoholic liver disease: update on the role of dietary fat. Biomol Ther 6 (1):1. https://doi.org/10.3390/biom6010001
19. Bertola A, Mathews S, Ki SH, Wang H, Gao B (2013) Mouse model of chronic and binge ethanol feeding (the NIAAA model). Nat Protoc 8(3):627–637. https://doi.org/10.1038/nprot.2013.032

# INDEX



Myriam Aouadi and Valerio Azzimato (eds.), *Kupffer Cells: Methods and Protocols*, Methods in Molecular Biology, vol. 2164, https://doi.org/10.1007/978-1-0716-0704-6, 

MIX
Papier aus verantwortungsvollen Quellen
Paper from responsible sources
FSC® C105338

If you have any concerns about our products,
you can contact us on
**ProductSafety@springernature.com**

In case Publisher is established outside the EU,
the EU authorized representative is:
**Springer Nature Customer Service Center GmbH**
**Europaplatz 3, 69115 Heidelberg, Germany**

Printed by Libri Plureos GmbH
in Hamburg, Germany